Phonocardiography
Signal Processing

Synthesis Lectures on Biomedical Engineering

Editor
John D. Enderle, *University of Connecticut*

Phonocardiography Signal Processing
Abbas K. Abbas and Rasha Bassam

ISBN: 978-3-031-00509-1   paperback
ISBN: 978-3-031-01637-0   ebook

DOI 10.1007/978-3-031-01637-0

A Publication in the Springer series
SYNTHESIS LECTURES ON BIOMEDICAL ENGINEERING

Lecture #31
Series Editor: John D. Enderle, University of Connecticut

Series ISSN
Synthesis Lectures on Biomedical Engineering
Print 1930-0328   Electronic 1930-0336

Phonocardiography
Signal Processing

Abbas K. Abbas
RWTH Aachen University

Rasha Bassam
Aachen University of Applied Science

SYNTHESIS LECTURES ON BIOMEDICAL ENGINEERING #31

ABSTRACT

The auscultation method is an important diagnostic indicator for hemodynamic anomalies. Heart sound classification and analysis play an important role in the auscultative diagnosis. The term phonocardiography refers to the tracing technique of heart sounds and the recording of cardiac acoustics vibration by means of microphone-transducer. Therefore, understanding the nature and source of this signal is important to give us a tendency for developing a competent tool for further analysis and processing, in order to enhance and optimize cardiac clinical diagnostic approach. This book gives the reader an inclusive view of the main aspects in phonocardiography signal processing.

KEYWORDS

phonocardiography, auscultation technique, signal processing, signal filtering, heart sounds, stethoscope microphone, cardiac acoustic modeling, wavelets analysis, data classification, spectral estimation and analysis, PCG classification, phonocardiography calibration, intracardiac phonocardiography, cardiac acoustic imaging

To our great land, Mesopotamia,
To our great Iraq,
To our light in the darkness: our parents.

Contents

Preface

In modern health care, auscultation has found its main role in personal health care and in decision making of particular and extensive clinical examination cases. Making a clinical decision based on auscultation is a double-edged sword: a simple tool which is able to screen and assess murmurs is imprecise, yet it would be both time- and cost-saving while also relieving many patients of needless apprehension. The instructions found in this book provide both a constructive and supportive background for students and biomedical engineers, because they provide not only the facts about the phonocardiography (PCG) signal but also the interaction between the heart sounds and PCG analysis platform, through advanced PCG signal processing methods. This approach will assist in identifying and obtaining useful clinical and physiological information. Although these PCG acquisition techniques are plain, noninvasive, low-cost, and precise for assessing a wide range of heart diseases, diagnosis by auscultation requires good experience and considerable observation ability.

The PCG signal is traditionally analyzed and characterized by morphological properties in time domain, by spectral properties in the frequency domain, or by non-stationary properties in a combined time-frequency domain. Besides reviewing these techniques, this book will cover recent advancement in nonlinear PCG signal analysis, which has been used to reconstruct the underlying cardiac acoustics model.

This processing step provides a geometrical interpretation of the signal's dynamics, whose structure can be used for both system characterization and classification, as well as for other signal processing tasks such as detection and prediction. In addition, it will provide a core of information and concepts necessary to develop modern aspects for an intelligent computerized cardiac auscultation, as a smart stethoscope module.

Particularly, this book will focus on classification and analysis of patterns resulting from PCG signal by using adaptive signal processing methods. System identification and modeling of main acoustic dynamics for a cardiac system based on different methods, such as autoregressive moving average ARMA, Páde approximation, and recursive estimation, are addressed. The PCG variation detection of several clinical-oriented diseases, such as mitral valve insufficiency and aortic regurgitation, is based on recurrence time statistics in combination with nonlinear prediction to remove obscuring heart sounds from respiratory sound recordings in healthy and patient subjects.

This book also lights up advanced aspects in the field of phonocardiography pattern classification and other higher-order data clustering algorithm. The application of artificial intelligence in PCG classification and data mining was discussed through artificial neural networks (e.g., Perceptron classifier and self-organized mapping (SOM) and fuzzy-based clustering method like fuzzy c-mean algorithm.

A special topic in PCG-related application was presented as fetal phonocardiography (fPCG) signal acquisition and analysis, PCG driven-rate responsive cardiac pacemaker, intracardiac phono-cardiography instrumentation and processing aspects, in addition to selected topics in physiological-derived signal with synchronization of PCG signal.

Finally, in this book, nonlinear PCG processing, as well as precise localization techniques of the first and second heart sound by means of ECG-gating method, are discussed and presented. Specifically learning objectives of each chapter will provide the students, physicians, and biomedical engineers with a good knowledge by introducing nonlinear analysis techniques based on dynamical systems theory to extract precise clinical information from the PCG signal.

Abbas K. Abbas and Rasha Bassam
Aachen, Germany
March 2009

List of Abbreviations

Abbreviation	Description
AV	Aortic valve
AVN	Atrioventricular node
AWD	Adaptive wavelet decomposition
ASD	Aortic stenosis disease
AHA	American Heart association
ARMA	Auto regressive moving average
ACOD	Audio Codecs
ACF	Auto –Correlation Function
ADC	Analogue-to-Digital Conversion
AKM	Adaptive K-mean Clustering Algorithm
ANN	Artificial neural network
BAD	Bradycardia arterial disease
BW	Bandwidth (of waveform)
BPF	Band pass filter
CAD	Congestive aortic disease
CM	Cardiac Microphone
CDS	Clinical Diagnosis System
CHD	Congestive heart disease
CWT	Continuous wavelet decomposition
DAQ	Data Acquitsion System
DbW	Debauchies wavelet
DCT	Discrete Cosine Transform
DFT	Discrete Fourier Transform
DS	Digital stethoscope
DVI	Pacemaker mode (Dual sensed, ventricular paced, inhibited mode)
DVT	Pacemaker mode (Dual sensed, ventricular paced, triggered mode)
DWT	Discrete wavelet decomposition
ECG	Electrocardiography
ePCG	Esophageal phonocardiography
ESD	Early systolic disease
ESPRIT	Estimation of Signal Parameters via Rotational Invariance Techniques

continues

Abbreviation	Description
continued	
FFT	Fast Fourier transform
FIR	Finite impulse response
FCM	Fuzzy C-mean classifier system
FHR	Fetal Heart rate
fPCG	Fetal phonocardiography
HBF	High pass filter
HMM	Hidden markov's model
HOS	Higher-order statistics
HT	Hilbert Transform
ICP	Intracardiac phonocardiography signal
ICSP	Intracardiac sound pressure signal
ICD	Intracardiac defibrillator
ICA	Independent Component Analysis
IEEE	Institute of Electrical and Electronic Engineering
IIR	Infinite impulse response filter
IABP	Intra-Aortic Balloon Pump
IRS	Image reconstruction system
KLM	Kalman linear model
LCD	Liquid crystal display
LVPV	Left ventricular Pressure Volume
LVP	Left ventricular Pressure
LVV	Left ventricular volume
LPF	Low pass filter
LTI	Linear-Time invariant system
LSE	Least-squares estimation
MSM	Mitral stenosis disease
MR	Mitral stenosis disease
MCU	Microcontroller unit
MI	Myocardial infarction
MV	Mitral Valve
OP	Operational point (Blood pressure curve)
ODE	Ordinary differential equation
PCA	Principal component Analysis
PCG	Phonocardiography
continues	

Abbreviation	Description
continued	
PDE	Partial differential equation
PET	Positron emission tomography imaging
PSG	Phonospirography signal
PATI	Phonocardiography acoustic tomography imaging
PSD	Power Spectral Density
P-wave	ECG cycle segment represent atrial depolarization phase
QRS-complex	ECG cycle segment represent ventricular depolarization phase
RBANN	Radial Basis Artificial Neural Network
RTF	Radial transformation
SOM	Self-Organized mapping
STFT	Short-time Fourier transform
SPECT	Single Photon Emission Computerized tomography imaging
SNR	Signal-to-noise ratio
SAN	Sino-Atrial node
S1	First heart sound
S2	Second heart sound
S3	Third heart sound
S4	Fourth heart sound
T-wave	ECG cycle segment represents ventricular repolarization phase
TAD	Tricuspid arterial disease
VVT	Pacemaker mode (ventricular sensed, ventricular paced, triggered mode)
VES	Visual electronic stethoscope
WAV	File format for audio-waveform data
WDE	Wavelet density estimation

List of Symbols

α	angle of fourier transformation
S_1	First heart sound
S_2	Second heart sound
S_3	Third heart sound
S_4	Fourth heart sound
θ_{PCG}	PCG pattern vector
V_s	microphone voltage source
R_s	microphone source impedance
ω	angular frequency
C_0	microphone output capacitance
V_0	microphone voltage output
Z_{ch}	acoustic impedance
A_2	Atrial component of PCG signal
P_2	Pulmonary component of PCG signal
$f(t)$	Fourier transform of PCG signal
E_t	PCG signal Energy
$\Phi(t)$	Haar Wavelet transform function
$\Psi(t)$	Haar scaling factor of PCG signal
Φ_2^D	Db-wavelet transformation of PCG signal
p_{2k}	two scale frequency-wavelet domain
Φ_{PCG}	Entropy value of PCG signal
$\gamma(s, \tau)$	continuous-wavelet transformation (CWT) of PCG signal
$\Psi(s, \tau)$	Scaling factor of (CWT) PCG signal
$S(t, w)$	Wavelet decomposition vector of PCG signal
$Mj - PCG(t)$	Spectral mean estimate of PCG signal
x_{PCG}	PCG signal data array
$\hat{R}_{PCG}(s)$	Power spectral density of PCG signal
$\hat{R}_B(w)$	PCG signal periodigram estimator
$H(z)$	Density transfer function of PCG signal
$A(z)$	Density transfer function zeros of PCG signal
$B(z)$	Density transfer function poles of PCG signal
X_{pcg}^T	PCG transfer matrix signal
f_{LO}	Microphone center frequency
w_f	Fundamental frequency

CHAPTER 1

Introduction to Phonocardiography Signal Processing

1.1 INTRODUCTION

Heart sounds result from the interplay of the dynamic events associated with the contraction and relaxation of the atria and ventricles, valve movements, and blood flow. They can be heard from the chest through a stethoscope, a device commonly used for screening and diagnosis in primary health care. The art of evaluating the acoustic properties of heart sounds and murmurs, including the intensity, frequency, duration, number, and quality of the sounds, are known as cardiac auscultation. Cardiac auscultation is one of the oldest means for assessing the heart condition, especially the function of heart valves. However, the traditional auscultation involves subjective judgment by the clinicians, which introduces variability in the perception and interpretation of the sounds, thereby affecting diagnostic accuracy.

With the assistance of electronic devices, phonocardiography (noninvasive technique)—a graphic recording of heart sounds—can be obtained, leading to more objective analysis and interpretation. In the earlier days, phonocardiography devices were used to document the timings and relative intensities of the components of heart sounds. However, they were generally inconvenient to use.

Further improvement in analog and digital microelectronics in the past decades has led to the development of the electronic stethoscope and its integrative functionality. These portable electronic stethoscopes allow clinicians to apply both auscultation and phonocardiography more conveniently. The new stethoscopes have also opened the possibilities for the application of advanced signal processing and data analysis techniques in the diagnosis of heart diseases. The practice of cardiac auscultation has come to a new era [1].

In the following chapters of this book, a focus on the biomedical engineering application of cardiac auscultation will considered, regarding the mechanical design, signal processing, data mining, clinical aided diagnosis, and medical standardization of this effective clinical technique.

Due to the growing field of dynamic biomedical signal modeling and system identification, additional mathematical analysis and modeling of stethoscope operation will be illustrated in a separated chapter. The different methods and a novel analysis algorithm for dynamic assessment of cardiac acoustics signal, such as PCG but not limited to, will improve the associated researchers for better understanding of PCG signal nature and its reflection on integrative clinical diagnosis of cardiomyopathy.

1.2   SIGNAL PROCESSING

What is signal processing? This question will be answered by carefully defining each of the words signal and the processing. Signal is a function of a set of independent variables, with time being perhaps the most prevalent single variable. The signal itself carries some kind of information available for observation. Processing is mean operating in some fashion on signal to extract some useful information. In many cases this processing will be a nondestructive "transformation" of the given data signal; however, some important processing methods turn out to be irreversible and thus destructive [2].

Our world is full in signals—some of these signals are natural, but most of the signals are man made. Some signals are necessary (speech), some are pleasant (music), while many are unwanted or unnecessary in a given situation. In an engineering context, signals are carriers of information, both useful and unwanted. Therefore, extracting or changing the useful information from a mix of conflicting information is the simplest form of signal processing. More generally, signal processing is an operation designed for extracting, enhancing, storing, and transmitting useful information.

The distinction between useful and unwanted information is often subjective as well as objective. Hence, signal processing tends to be application dependent [3, 4].

1.2.1   OVERVIEW OF SIGNAL PROCESSING

Originally, signal processing was done only on analog or continuous time signals using analog signals processing (ASP). Until the late 1950s digital computers were not commercially available. When they did become commercially available they were large and expensive, and they were used to simulate the performance of analog signal processing to judge its effectiveness. These simulations, however, led to digital processor code that simulated or performed nearly the same task on samples of the signals that the analog simulation coding of the analog system was actually a digital signal processing (DSP) system that worked on samples of the input and output at discrete time intervals. But implementing signal processing digitally instead of using analog systems was still out of the question.

The first problem was that an analog input signal had to be represented as a sequence of samples of the signal, which were then converted to the computer's numerical representation. The same process would have to be applied in reverse to the output of the digitally processed signal. The second problem was that because the processing was done on very large, slow, and expensive computers, practical real-time processing between samples of the signal was impossible. The signals that we encounter in practice are mostly analog signals. These signals, which vary continuously in time and amplitude, are processed using electrical networks containing active and passive circuit elements. This approach is known as analog signal processing (ASP). They can also be processed using digital hardware, however, one needs to convert analog signals into a form suitable for digital hardware: this form of the signal is called a digital signal and it takes one of the finite numbers of values at specific instances in time, and hence, it can be represented by binary numbers, or bits.

The processing of digital signals is called DSP [3].

Two conceptual schemes for the processing of signals are illustrated in Fig. 1.1. The digital processing of analog signals requires an analog-to-digital converter (ADC) for sampling the analog signal and a digital-to-analog converter (DAC) to convert the processed digital signal back to analog form [4].

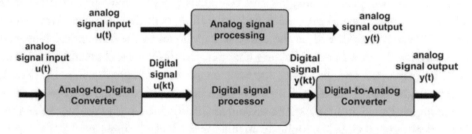

Figure 1.1: Analog and digital signal processing concepts.

It appears from the above two approaches to signal processing, analog and digital, that the DSP approach is the more complicated, containing more components than the ASP. Therefore, one might ask a question: Why process signals digitally? The answer lies in many advantages offered by DSP. Some of the advantages of a DSP system over analog circuitry are summarized as follows [5]:

- Flexibility. Function of DSP system can be easily modified and upgraded with software that has implemented the specific algorithm for using the same hardware. One can design a DSP system that can be programmed to perform a wide variety of tasks by executing different software modules.

- Reproducibility. The performance of a DSP system can be repeated precisely from one unit to another. This is because the signal processing of DSP system works directly with binary sequences.

- Reliability. The memory and logic of DSP hardware does not deteriorate with age. Therefore, the field performance of DSP systems will not drift with changing environmental conditions or aged electronic components as their analog counterparts do.

- Complexity. Using DSP allows sophisticated applications such as speech or image recognition to implement for lightweight and low-power portable devices. This is impractical using traditional analog techniques

With the rapid evolution in technology in the past several years, DSP systems have a lower overall coast compared to analog systems. The principal disadvantage of DSP is the speed of operations, especially at very high frequencies.

Primarily due to the above advantages, DSP is now becoming a first choice in many technologies and applications, such as consumer electronics, communications, wireless telephones, and medical engineering.

1.3 APPLICATION OF SIGNAL PROCESSING IN BIOMEDICAL ENGINEERING

Several review articles on medical imaging and biosignals [6]–[17] have provided a detailed description of the mainstream signal processing functions along with their associated implementation considerations. These functions will be the effective techniques in the biomedical-biosignals processing and analysis schemes. In the following chapters, the different and several signal processing methods will be presented and discussed, focusing on phonocardiography signal processing and higher-order analysis, such as (classification, data clustering, statistical signal processing, and cardiac acoustic modeling and identification) will be presented. As Fig. 1.2 displays the block diagram of the general biomedical signal processing application, this approach can be integrated as a computational core for computer aided diagnosis system.

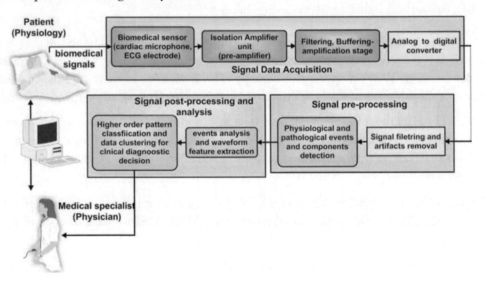

Figure 1.2: Block diagram of the general biomedical signal processing and analysis, as an integrative approach for computer-aided diagnosis system.

1.4 CARDIOVASCULAR PHYSIOLOGY

The heart is one of the most important organs of the human body. It is responsible for pumping deoxygenated blood to the lungs, where carbon dioxide-oxygen (CO_2-O_2) exchange takes place, and pumping oxygenated blood throughout the body. Anatomically, the heart is divided into two sides: the left side and the right side, which are separated by the septum. Each side is further divided into two chambers: the atrium and the ventricle.

As illustrated in Fig. 1.3 [19, 20], heart valves exist between the atria and the ventricles and between the ventricles and the major arteries from the heart, which permit blood flow only in one direction. Such valves include the tricuspid valve, the mitral valve, the pulmonary valve, and the aortic valve. The tricuspid and mitral valves are often collectively called the atrioventricular valves, since they direct blood flow from the atria to the ventricles. The competence of the atrioventricular valves depends not only on the proper functioning of the valve leaflets themselves but also on the strong fibrous strands, called chordate tendineae, which are attached to the free edges and ventricular surfaces of the valve cusps. These strands are, in turn, attached to the finger-like projections of the muscle tissue from the endocardium called the papillary muscles [19].

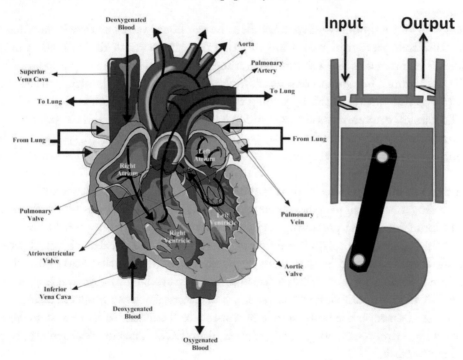

Figure 1.3: Right side: vertical section of the cardiac muscle shows the internal structure of the heart. Left side: schematic representation of a reciprocating type pump having a pumping chamber and input output ports with oppositely oriented valves.

The heart can be classified from a hemodynamics point of view as a simple reciprocating pump. The mechanical principles of a reciprocating pump are illustrated in Fig. 1.3. The pumping chambers have a variable volume and input and output ports. A one-way valve in the input port is oriented such that it opens only when the pressure in the input chamber exceeds the pressure within the pumping chamber. Another one-way valve in the output port opens only when pressure in the pumping chamber exceeds the pressure in the output chamber. The rod and crankshaft will cause the diaphragm to move back and forth. The chamber's volume changes as the piston moves, causing the pressure within to rise and fall. In the heart, the change in volume is the result of contraction and relaxation of the cardiac muscle that makes up the ventricular walls.

One complete rotation of the crankshaft will result in one pump cycle. Each cycle, in turn, consists of a filling phase and an ejection phase. The filling phase occurs as the pumping chamber's volume is increasing and drawing fluid through the input port. During the ejection phase, the pumping chamber's volume is decreasing and fluid is ejected through the output port. The volume of fluid ejected during one pump cycle is referred as the stroke volume and the fluid volume pumped each minute can be determined by simply multiplying the stroke volume times the number of pump cycles per minute.

The aortic and pulmonary valves are called the semilunar valves as they have a half-moon-shaped structure that prevents the backflow of blood from the aorta or the pulmonary artery into the ventricles. The heart acts like a pump, generating the required pressure to pump blood through the arterial circulation. The process consists of synchronized activities of the atria and the ventricles. First, the atria contract (atria systole) pumping the blood into the ventricles. As the atria begin to relax (atrial diastole), the ventricles contract to force blood into the aorta and the pulmonary artery (ventricular systole). Then the ventricles relax (ventricular diastole). During this phase, both the atria and the ventricles relax until atrial systole occurs again. The entire process is known as the cardiac cycle.

The diagram shown in Fig. 1.4 consists of three main stages: (1) signal data acquisition, (2) signal pre-processing, and (3) signal post-processing and analysis, in which this block diagram shows the general biomedical signal processing and analysis, as an integrative approache for computer-aided diagnosis. Figure 1.4 displays the different interactions of the cardiac electrical activity, the inter-dynamics between the different systems involved, and the various electrical signals that represent the different cardiac activities. The electrical conduction system is the main rate controller, and it is regulated by the autonomous nervous system. The electrical activity results in action potentials that are conducted through the heart muscle by a specialized conductive tissue system, and can be measured as voltage differences on the body surface, using ECG. The electrical activity triggers the mechanical contraction.

The mechanical activity of the heart involves contraction of myocardial cells, opening/closing of valves, and flow of blood to and from the heart chambers. This activity is modulated by changes in the contractility of the heart, the compliance of the chamber walls and arteries and the developed pressure gradients. The mechanical activity can be also examined using ultrasound imaging.

The peripheral blood flows in the arteries and veins is also modulated by mechanical properties of the tissue. The flow of blood can be imaged by Doppler-echo, and the pulse-wave can be captured in one of the peripheral arteries. The different types of signals give us various pieces of information

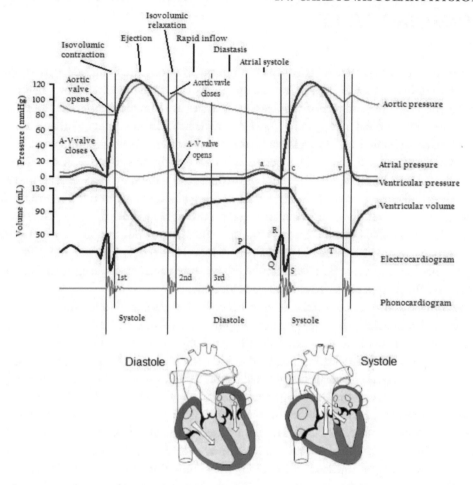

Figure 1.4: Cardiac cycle events occurring in the left ventricle. Above: Pressure profile of the ventricle and atrium. Middle: Volume profile of the left ventricle. Below: Phonocardiographgy signals. This diagram consists of three main stages: (1) signal data acquisition, (2) signal pre-processing, and (3) signal post-processing and analysis.

about the cardiac activity. Integrating this information may yield a better ability to assess the condition of the cardiovascular system. The detailed events in the cardiac cycle will explained in the following section.

1.5   CARDIAC CYCLE

The heart is actually composed of two separate pumps, one on the right side that supplies the pulmonary circulation and one on the left that supplies the systemic circulation. The principles that regulate the flow into and out of the heart's ventricles are somehow different from the action of the illustrated mechanical pump, as in Fig. 1.3, which has a fixed stroke volume. The temporal relationships between ventricular contraction and blood flow in the heart are illustrated in Fig. 1.3. When the ventricular muscle is relaxed, a period referred to as diastole, the pressure in the ventricle will be less than the pressures within the veins and atria, because the ventricle will relax under closed semilunar valves and closed atreoventricular valves, causing blood to flow into the ventricles through the atrioventricular (mitral on the left and tricuspid on the right) valves.

The relaxed ventricle cannot create a negative pressure to pull blood into it. Instead, the ventricular lumen can only be distended passively with blood under a positive pressure. That pressure must be generated in the veins that feed the heart. Because ventricular filling is in proportion to venous pressure, the heart's stroke volume is quite variable.

After the end of diastole, the atria will start to contract to push the blood through the atreoventricular valve to the ventricle, and because there are no valves between the atria and the veins, much of the atrial blood is actually forced back into the veins. Nevertheless, atrial contraction will push additional blood into the ventricles, causing further increases in ventricular pressure and volume. Although the benefit of atrial contraction at normal resting condition of the body may be negligible, it can substantially increase ventricular filling at exercise and high heart rates when diastolic filling time is curtailed or the need for extra cardiac output is needed. As the ventricular musculature contracts, a period termed systole, the force in the walls is transmitted to the blood within the ventricular lumen. Ventricular pressure increases and as it rises above atrial pressure, so the atrioventricular valves will close. The heart now begins a period of isovolumetric contraction as pressure builds in the lumen. No blood can enter or leave the ventricle because both the inflow and the outflow valves are closed. When pressure in the ventricular lumen finally exceeds that in the outflow vessel (the aorta for the left heart and the pulmonary artery for the right heart), the semilunar valves (aortic on the left and pulmonary on the right) will be opened and blood is ejected.

The heart now begins a period of isovolumetric contraction as pressure builds in the lumen. No blood can enter or leave the ventricle because both the inflow and the outflow valves are closed. When pressure in the ventricular lumen finally exceeds that in the outflow vessel (the aorta for the left heart and the pulmonary artery for the right heart), the semilunar valves (aortic on the left and pulmonary on the right) open and blood is ejected.

As systole ends, the ventricular musculature relaxes and the force exerted on the blood in the ventricular lumen subsides. Ventricular pressure falls below outflow pressure in the outflow vessel and the semilunar valves close. At this point, both the semilunar and the atrioventricular valves are closed so that a second isovolumetric period occurs. Atrial blood will not flow into the ventricles until relaxation has proceeded to the point when ventricular pressure falls below atrial pressure. When that occurs, the atrioventricular (AV) valves open and the filling phase of the cardiac cycle once again repeats itself. [1].

1.6   CARDIAC PRESSURE PROFILE

The physician can best appreciate the events of the cardiac cycle by measuring the pressures at various locations in the cardiovascular system with a catheter. Cardiac catheterization has become a powerful tool for the diagnosis of cardiac disease and the student must, therefore, become thoroughly familiar with the pressure profiles in the atria, ventricles, and great vessels. Formally, seven distinct phases during a single cardiac cycle are recognized. Figure 1.4 illustrates how aortic pressure, left ventricular pressure, left atrial pressure, left ventricular volume, and the ECG are temporally correlated throughout these seven phases.

Period A in Fig. 1.5 represents atrial systole. Note that contraction of the left atrium causes both the left ventricular and left atrial pressure to rise by a few mmHg. This rise in the atrial pressure is called the A wave. As the atrium begins to relax, atrial pressure falls causing the X wave. The volume of blood present in the ventricle at the end of atrial systole is termed the end-diastolic volume. In period B, the isovolumetric period of contraction, ventricular pressure is seen to separate from atrial pressure because of closure of the mitral valve. The upward movement of the mitral valve into the atrium causes the C wave. This is followed by a second fall in atrial pressure, the X_0 wave.

[1]Medical auscultation technique applied to any acoustic measurement in human body

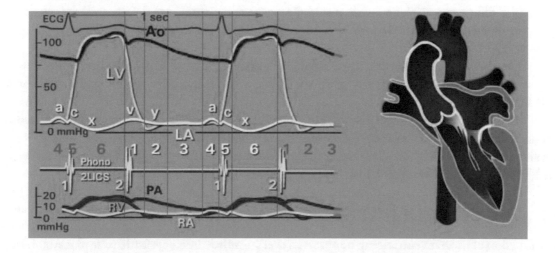

Figure 1.5: Phonocardiography synchronization with hemodynamic tracing in cardiac cycle showing the fundamental events of this cycle and the associated electrical, mechanical, acoustic annotation, and pressure waveform for corresponding cycle events.

The isovolumetric period ends as left ventricular pressure reaches arterial pressure and the aortic valve opens. During period C, most of the stroke volume is ejected into the aorta, as shown by the volume trace; hence, the term rapid ejection phase. The next phase, D, is termed the reduced ejection period. During both ejection periods, the aortic valve opens, making the aorta and left ventricle a common chamber and the pressure within them nearly equal.

During rapid ejection the velocity at which blood is being ejected is increasing, causing ventricular pressure to slightly lead that in the aorta by a few mmHg. As the rate of ejection slows during the reduced ejection period, the inertia of the decelerating column of blood traveling down the aorta reverses the gradient causing aortic pressure to slightly lead ventricular pressure. As the ventricle begins to relax, pressure in the ventricle falls. As blood begins to flow backward across the aortic valve, it closes its leaflets. That momentary retrograde flow of blood at the aortic valve and its abrupt deceleration as the valve snaps closed cause a small rebound in the aortic pressure trace called the dicrotic notch.

The volume of blood left in the ventricle at aortic valve closure is termed the end-systolic volume. During the isovolumetric period of relaxation, E, left ventricular and aortic pressure separate and ventricular pressure continues to fall. The isovolumetric relaxation period ends when ventricular pressure reaches below the left atrial pressure and the mitral valve opens. Although the mitral valve is closed during ventricular systole, ventricular contraction causes bulging of the atreoventricular valve into the atria and cause its pressure to rise slightly, generating the V wave in the atrial pressure tracing. This elevated pressure causes blood to surge into the ventricle as soon as the mitral valve opens. For that reason, period F is called the rapid filling phase. The abrupt fall in atrial pressure during the rapid filling phase gives rise to the Y wave. During the remainder of diastole, the reduced ventricular filling period, the pressure within the ventricle has equilibrated with atrial pressure, and little additional blood enters the ventricle. As atrial blood fills the ventricle, atrial pressure rises once more as the H wave.

The pressure in the aorta is the arterial blood pressure. The peak pressure during ejection is referred to as the systolic pressure, whereas the lowest pressure just prior to aortic valve opening is called the diastolic pressure.

Since the diagram in Figs. 1.4 and Fig. 1.5 is for the illustration of cardiac cycle events in the left heart and the aortic zone, the pressure relationships within the right heart and pulmonary artery are also illustrated in Fig. 1.5 (lower part of the figure) and they are virtually identical to those of the left heart, with the exception that the pressures are only about one-fifth as great.

1.7   VENTRICULAR PRESSURE-VOLUME LOOPS

The function of the left ventricle can be observed over an entire cardiac cycle (diastole plus systole) by combining the two pressure-volume relationships. By connecting these two pressure-volume curves, it is possible to construct a so-called ventricular pressure-volume loop (Fig. 1.6). Recall that the systolic pressure-volume relationship in Fig. 1.5 shows the maximum developed ventricular pressure for a given ventricular volume.

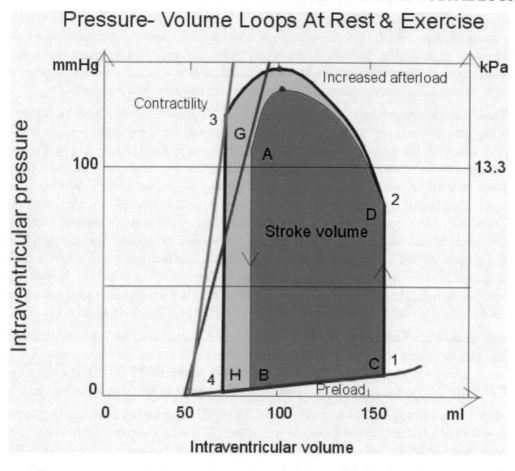

Figure 1.6: Schematic diagram of left ventricular pressure volume loop.

To facilitate understanding, a portion of that systolic pressure-volume curve is superimposed as a gold line on the ventricular pressure-volume loop. The line shows the maximum possible pressure that can be developed for a given ventricular volume during systole, i.e., when the ventricle is contracting. Note that point (3) in Fig. 1.5 on the pressure-volume loop touches the systolic pressure-volume curve. Also, it may not be evident that the portion of the loop between points 4 and 1 corresponds to a portion of the diastolic pressure-volume curve. The ventricular pressure-volume loop shown in Fig. 1.6 describes one complete cycle of ventricular contraction, ejection, relaxation, and refilling as follows:

- **Isovolumetric contraction phase** (1→2). Begin the cycle at point 1, which marks the end of diastole. The left ventricle has filled with blood from the left atrium, and its volume is the end-diastolic volume, 140 mL. The corresponding pressure is quite low because the ventricular muscle is relaxed. At this point, the ventricle is activated, it contracts, and ventricular pressure increases dramatically. Because all valves are closed, no blood can be ejected from the left ventricle, and ventricular volume is constant, although ventricular pressure becomes quite high at point 2. Thus, this phase of the cycle is called isovolumetric contraction.

- **Ventricular ejection phase** $(2 \rightarrow 3)$. At point 2, left ventricular pressure becomes higher than aortic pressure, causing the aortic valve to open. (You may wonder why the pressure at point 2 does not reach the systolic pressure-volume curve shown by the dashed gold line. The simple reason is that it does not have to. The pressure at point 2 is determined by aortic pressure. Once ventricular pressure reaches the value of aortic pressure, the aortic valve opens and the rest of the contraction is used for ejection of the stroke volume through the open aortic valve.) Once the valve is open, blood is rapidly ejected, driven by the pressure gradient between the left ventricle and the aorta. During this phase, left ventricular pressure remains high because the ventricle is still contracting. Ventricular volume decreases dramatically, however, as blood is ejected into the aorta. The volume remaining in the ventricle at point 3 is the end-systolic volume, 70 mL. The width of the pressure-volume loop is the volume of blood ejected, or the stroke volume. The stroke volume in this ventricular cycle is 70 mL (140-70 mL).

- **Isovolumetric relaxation phase** $(3 \rightarrow 4)$. At point 3, systole ends and the ventricle relaxes. Ventricular pressure decreases below aortic pressure and the aortic valve closes. Although ventricular pressure decreases rapidly during this phase, ventricular volume remains constant (isovolumetric) at the end-systolic value of 70 mL because all valves are closed again.

- **Ventricular filling phase** $(4 \rightarrow 1)$. At point 4, ventricular pressure has fallen to a level that now is less than left atrial pressure, causing the mitral (AV) valve to open. The left ventricle fills with blood from the left atrium passively and also actively, as a result of atrial contraction in the next cycle. Left ventricular volume increases back to the end-diastolic volume of 140 mL. During this last phase, the ventricular muscle is relaxed, and pressure increases only slightly as the compliant ventricle fills with blood.

Ventricular pressure-volume loops can be used to visualize the effects of changes in preload (i.e., changes in venous return or end-diastolic volume), changes in afterload (i.e., changes in aortic pressure), or changes in contractility.

1.8   CARDIAC ELECTRICAL CONDUCTION SYSTEM

The periodic activity of the heart is controlled by an electrical conducting system. The electrical signal originates in specialized pacemaker cells in the right atrium (the sino-atria node), and is propagated through the atria to the AV-node (a delay junction) and to the ventricles. The main events in generating and propagating the bio-action potential of the cardiac tissue are illustrated in Fig. 1.7. The electrical action potential excites the muscle cells and causes the mechanical contraction of the heart chambers.

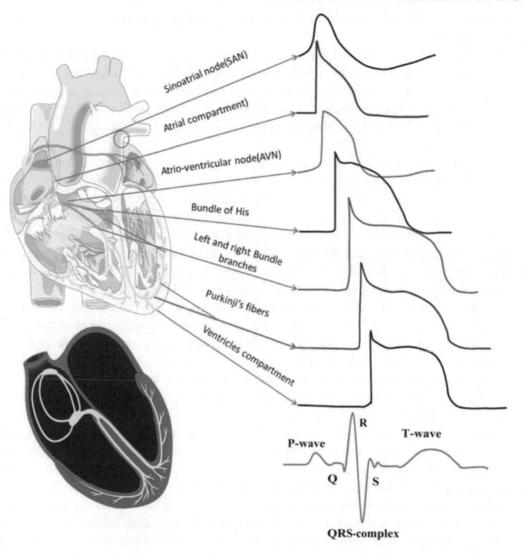

Figure 1.7: Right side: Cardiac electrical conduction system morphology and timing of action potentials from different regions of the heart. Left side: related ECG signal as measured on the body surface.

1.9   PHYSIOLOGY OF THE HEART SOUND

During the systolic and the diastolic phase of the cardiac cycle, audible sounds are produced from the opening and the closing of the heart valves, the flow of blood in the heart, and the vibration of heart muscles. Usually, four heart sounds are generated in a cardiac cycle. The first heart sound and the second heart sound can be easily heard in a normal heart through a stethoscope placed on a proper area on the chest. The normal third heart sound is audible in children and adolescents but not in most adults.

The fourth heart sound is seldom audible in normal individuals through the conventional mechanical stethoscopes but can be detected by sensors with high sensitivity, such as electronic stethoscopes and phonocardiography systems. Sounds other than these four, called murmurs, are abnormal sounds resulting from valve problems, or sounds made by artificial pacemakers or prosthetic valves. See Fig. 1.4 [21] for an illustration of how the four heart sounds are correlated to the electrical and mechanical events of the cardiac cycle.

The first heart sound (S_1) occurs at the onset of ventricular systole. It can be most clearly heard at the apex and the fourth intercostal spaces along the left sternal border. It is characterized by higher amplitude and longer duration in comparison with other heart sounds. It has two major high-frequency components that can be easily heard at bedside. Although controversy exists regarding the mechanism of S_1 [3], the most compelling evidence indicates that the components result from the closure of the mitral and tricuspid valves and the vibrations set up in the valve cusps, chordate, papillary, muscles, and ventricular walls before aortic ejection [21]. S_1 lasts for an average period of 100–200ms. Its frequency components lie in the range of 10-200 Hz.

The acoustic properties of S_1 are able to reveal the strength of the myocardial systole and the status of the atrioventricular valves' function. As a result of the asynchronous closure of the tricuspid and mitral valves, the two components of S_1 are often separated by a time delay of 20-30 ms. This delay is known as the (split) in the medical community and is of significant diagnostic importance. An abnormally large splitting is often a sign of heart problem. The second heart sound (S_2) occurs within a short period once the ventricular diastole starts. It coincides with the completion of the T-wave of the electrocardiogram (ECG).

These sounds can be clearly heard when the stethoscope is firmly applied against the skin at the second or third intercostals space along the left sternal border. S_2 consists of two high-frequency components, one because of the closure of the aortic valve and the other because of the closure of the pulmonary valve. At the onset of ventricular diastole, the systolic ejection into the aorta and the pulmonary artery declines and the rising pressure in these vessels exceeds the pressure in the respective ventricles, thus reversing the flow and causing the closure of their valves. The second heart sound usually has higher-frequency components as compared with the first heart sound. As a result of the higher pressure in the aorta compared with the pulmonary artery, the aortic valve tends to close before the pulmonary valve, so the second heart sound may have an audible split.

In normal individuals, respiratory variations exist in the splitting of S_2. During expiration phase, the interval between the two components is small (less than 30 ms). However, during inspiration, the splitting of the two components is evident. Clinical evaluation of the second heart sound is a bedside technique that is considered to be a most valuable screening test for heart disease. Many heart diseases are associated with the characteristic changes in the intensities of or the time relation between the two components of S_2. The ability to estimate these changes offers important diagnostic clues [21, 22]. S_1 and S_2 were basically the main two heart sounds that were used for most of the clinical assessment based on the phonocardiography auscultation procedure.

The third and fourth heart sounds, also called gallop sounds, are low-frequency sounds occurring in early and late diastole, respectively, under highly variable physiological and pathological conditions. Deceleration of mitral flow by ventricular walls may represent a key mechanism in the genesis of both sounds [21, 23]. The third heart sound (S_3) occurs in the rapid filling period of early diastole. It is produced by vibrations of the ventricular walls when suddenly distended by the rush of inflow resulting from the pressure difference between ventricles and atria. The audibility of S_3 may be physiological in young people or in some adults, but it is pathological in people with congestive heart failure or ventricular dilatation.

The presence of the third heart sound in patients with valvular heart disease is often regarded as a sign of heart failure, but it also depends on the type of valvular disease [22]. In patients with mitral regurgitation, the third heart sound is common but does not necessarily reflect left ventricular systolic dysfunction or increased filling pressure. In patients with aortic stenosis, third heart sounds are uncommon but usually indicate the presence of systolic dysfunction and elevated filling pressure. The fourth heart sound (S_4) occurs in late diastole and just before S_1. It is produced by vibrations in expanding ventricles when atria contract. Thus, S_4 is rarely heard in a normal heart. The abnormally audible S_4 results from the reduced distensibility of one or both ventricles. As a result of the stiff ventricles, the force of atrial contraction increases, causing sharp movement of the ventricular wall and the emission of a prominent S_4 [22, 24]. Most murmurs are the result of turbulent blood flow, which produces a series of vibrations in the cardiac structure. Murmurs during the early systolic phase are common in children, and they are normally heard in nearly all adults after exercise. Abnormal murmurs may be caused by stenosis and insufficiencies (leaks) at the aortic, pulmonary, or mitral valves [23]. It is important from a diagnostic point of view to note the time and the location of murmurs. The identification of murmurs may assist the diagnosis of heart defections like aortic stenosis, mitral and tricuspid regurgitation, etc. [2]

1.10 ABNORMAL HEART SOUND PATTERN

Dealing with the cardiac sound abnormalities (in Fig. 1.8 and Fig. 1.9), there are many cardiac defects that can lead to the production of additional (excessive) or modified heart sounds. Some of these abnormalities will be presented below. There are many web sites where one can download and listen to electronically recorded heart sounds as *.wav files. Also, in Canada, McGill University's Physiology and Music Departments have an unique Medical Informatics web site at which the viewer can listen to various cardiac sounds (normal and abnormal) at different chest recording sites. In addition, the viewer can download 3-D, colored-mesh, time-frequency spectrograms covering several cardiac cycles of a particular sound, as well as read text about the source and significance of the sound (Glass, 1997).

---

[2]Varieties of patho-physiological case have distinct heart murmurs.

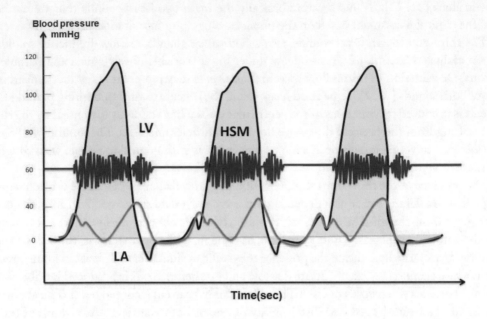

Figure 1.8: Pressure diagram of left ventricle (LV) and left atrium (LA) versus heart sound trace showing related heart sounds mummers in the case of aortic stenosis (AS) disorder.

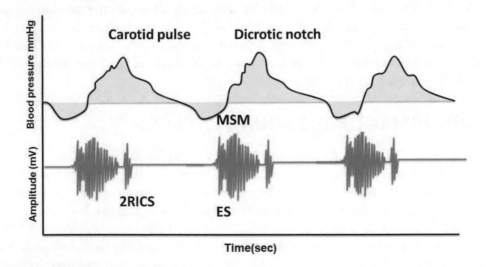

Figure 1.9: Cardiac cycle events for mitral valve stenosis disorder. Above: the carotid artery pulse. Below: mitral stenosis murmurs (MSM) and early systolic murmur (ES).

A major source of anomalous heart sounds is damaged heart valves. Heart valves, in particular the left heart valves, can either fail to open properly (they are stenosed) or they cannot close properly (they are incompetent), causing a backflow of blood, or regurgitation. A major source of heart valve damage can be infection by a hemolytic streptococcus, such as scarlet fever, sore throat, or middle ear infection. A serious complication is rheumatic fever, one of the characteristics of which is carditis and valvular damage. The streptococcus bacteria manufacture a protein called the (M antigen) to which the immune system forms antibodies.

Unfortunately, these antibodies also attack certain body tissues, notably the joints and the heart. Guyton (1991) states: "In rheumatic fever, large hemorrhagic, fibrinous, bulbous lesions grow along the inflamed edges of the heart valves." The scarring from this autoimmune inflammation leaves permanent valve damage. The valves of the left heart (aortic and mitral) are the most prone to damage by antibodies.

Table 1.1 shows the principle characteristics of the heart mummers which derived spatially and have clinical importance in diagnosis in heart valve abnormalities and describes the different heart sound and the origin of each one. In aortic valve stenosis, the valve cannot open properly; there is an

Table 1.1: Spatial characteristics of diagnosing valve disease from heart murmurs		
Heart	Sound occurs during	Associated with
S_1	isovolumetric contraction	mitral and tricuspid valves closure
S_2	isovolumetric relaxation	aortic and pulmonary valves closure
S_3	early ventricular filling	normal in children; in adults, associated with ventricular dilation (e.g., ventricular systolic failure)
S_4	atrial contraction	associated with stiff, low compliant ventricle (e.g., ventricular hypertrophy)

abnormally high hydraulic resistance against which the left ventricle must pump. Thus, the peak left ventricular pressure can rise as high as 300 mmHg, while the aortic pressure remains in the normal range. The exiting blood is forced through the small aperture at very high velocity, causing turbulence and enhanced vibration of the root of the aorta. This vibration causes a loud murmur during systole that is characteristic of aortic stenosis. Aortic regurgitation, on the other hand, occurs because the damaged aortic valve does not close completely. Fig. 1.11 shows a schematic representation of typical cardiac variables: the ECG, the logic states of the heart valves, low- and high-frequency phonocardiograms, a recording of a vessel pulse (carotid artery), and of the heart apex pulse (apex-cardiogram). The heart cycle is divided into specific intervals according to the valve states of the left heart. The left ventricular systole is composed of the isovolumetric contraction and the ejection period; the left ventricular diastole covers the isovolumetric relaxation and the left ventricular filling (successively, the rapid filling, the slow filling, and the atrial contraction). A similar figure could be given for the right heart; valve phenomena are approximately synchronous with those of the left heart. Small time shifts are typical: mitral valve closure precedes tricuspid closure and aortic valve closure precedes pulmonary closure. The low-frequency PCG shows the four normal heart sounds (I, II, III, and IV). In the high frequency, trace III and IV have disappeared and splitting is visible in I and in II.

Again, there is a high-velocity jet of blood forced back into the left ventricle by aortic back-pressure during diastole (when the left ventricle is relaxed). This back-pressure makes it difficult for the left atrium to fill the left ventricle, and, of course, the heart must work harder to pump a given volume of blood into the aorta. The aortic regurgitation murmur is also of relatively high pitch, and has a swishing quality (Guyton, 2005, [20]).

In mitral valve stenosis, Fig. 1.9, the murmur occurs in the last two thirds of diastole, caused by blood's jetting through the valve from the left atrium to the left ventricle. Because of the low peak pressure in the left atrium, a weak, very low-frequency sound is produced. The mitral stenotic murmur often cannot be heard; its vibration must be felt, or seen on an oscilloscope from a microphone output. Another audible clue to mitral stenosis is an opening snap of the mitral valve, closely following the normal S_2.

Mitral valve regurgitation takes place during systole. As the left ventricle contacts, it forces a high-velocity jet of blood back through the mitral valve, making the walls of the left atrium vibrate. The frequencies and amplitude of mitral valve regurgitation murmur are lower than aortic valve stenosis murmur because the left atrium is not as resonant as the root of the aorta. Also, the sound has farther to travel from the left atrium to the front of the chest. Another cardiac defect that can be diagnosed by hearing the S_2 sound (split) is a left or right bundle branch block. The synchronization of the contraction of the muscle of the left and right ventricles is accomplished by the wave of electrical depolarization that propagates from the AV node, down the bundle of His, which bifurcates into the left and right bundle branches that run down on each side of the ventricular septum. Near the apex of the heart, the bundle branches branch extensively into the Purkinje fibers, which invade the inner ventricular cardiac muscle syncytium, carrying the electrical activity that triggers ventricular contraction. See Fig. 1.11 for a time-domain schematic of where certain heart sounds occur in the cardiac cycle. If the bundle branch fibers on the right side of the septum are damaged by myocardial infarction, the contraction of the right ventricle will lag that of the right, and the sound associated with the aortic valve's closing will lead that sound caused by the pulmonary valve. This split in sound S_2 is heard regardless of the state of inhale or exhale. A left bundle branch block will delay the contraction of the left ventricle, hence delay the aortic valve sound with respect to that of the pulmonary valve. This condition causes reverse splitting of S_2 during expiration, but is absent on inspiration. Other causes of the reverse split include premature right ventricular contraction (as opposed to a delayed left ventricular systole), or systemic hypertension (high venous return pressure).

1.10.1 HEART SOUND AS HEMODYNAMIC INDEX

There is a direct relation between the intensity and the frequency of sound II and the slope of ventricular pressure falling during its volumetric relaxation. Stiffening of the valve leaflets results in a reduction of sound II. A higher valve radius or a lowered blood viscosity gives rise to an increased second sound. Cardiovascular pathologies can have an effect on timing and intensities of the second heart sound components. Wide splitting of sound II can be due to delayed pulmonary valve closure or advanced aortic valve closure. Delayed pulmonary valve closure can be caused by right bundle branch block, pulmonary stenosis, pulmonary hypertension, and atrial septum defect; advanced aortic valve closure can result from mitral regurgitation and ventricular septum defect. Paradoxical splitting of sound II can be due to delayed aortic valve closure or advanced pulmonary valve closure.

Delayed aortic valve closure can be caused by left bundle branch block, aortic stenosis, and arteriosclerotic heart disease. Advanced pulmonary valve closure can be caused by tricuspid regurgitation and advanced right ventricular activation. IIA and IIP, respectively, can be absent in severe aortic and pulmonary valve stenosis. IIA is decreased in aortic regurgitation and in pathologically diminished left ventricular performance. The third sound (III) occurs during the rapid passive filling period of the ventricle. It is believed that III is initiated by the sudden deceleration of blood flow when the ventricle reaches its limit of distensibility, causing vibrations of the ventricular wall. It can often be heard in normal children and adolescents, but can also be registered in adults (although not heard) in the low-frequency channel. It is a weak and low-pitched (low-frequency) sound. Disappearance of III is a result of aging as a consequence of increasing myocardial mass having a damping effect on vibrations. High filling rate or altered physical properties of the ventricle may cause an increased third sound. If III reappears with aging (beyond the age of 40 years), it is pathological in most cases. A pathological sound III is found in mitral regurgitation, aortic stenosis, and ischemic heart disease. The fourth sound (IV) coincides with the atrial contraction and thus the originated increased blood flow through the mitral valve with consequences as mentioned for the third sound. It is seldom heard in normal cases, sometimes in older people, but is registered more often in the low-frequency channel. The sound is increased in cases of augmented ventricular filling or reduced ventricular distensibility.

A pathological sound IV is found in mitral regurgitation, aortic stenosis, hypertensive cardiovascular disease, and ischemic heart disease. Besides these four sounds, some pathological heart sounds may be present. Among the systolic sounds there is the ejection sound and the non-ejection systolic click. The ejection sound can be found in different pathological conditions such as congenital aortic or pulmonary valvular stenosis where opening of the cusps is restricted. A non-ejection systolic click may be associated with a sudden mitral valve prolapsed into the left atrium. An opening snap, a diastolic sound, may occur at the time of the opening of the mitral valve, for example, in cases with valve stenosis.

Heart murmurs are assumed to be caused by different mechanisms as compared to heart sounds. In fact, most murmurs result from turbulence in blood flow and occur as random signals. In normal blood vessels at normal velocity values blood flow is laminar, that is, in layers, and no turbulence is observed. In a normal resting human, there may be turbulent flow only in the vicinity of the aortic and pulmonary valves. As flow turbulence, a phenomenon that is generally irregular and random, is associated with pressure turbulence and, consequently, vessel wall vibration, acoustic phenomena may be observed. For flow in a smooth straight tube, the value of the Reynolds number, a dimensionless hydrodynamic parameter, determines the occurrence of turbulence. This number is proportional to the flow velocity and the tube diameter, and inversely proportional to the viscosity of the fluid. If this number exceeds a threshold value, laminar flow becomes turbulent.

According to this theory, so-called innocent murmurs can be explained: They are produced if cardiac output is raised or when blood viscosity is lowered; they are generally early or mid-systolic, have a short duration, and coincide with maximum ventricular outflow. Turbulence, and thus intensity, of the murmur increase with flow velocity.

Pathological murmurs may be originated at normal flow rate through a restricted or irregular valve opening (e.g., in cases of valve stenosis) or by an abnormal flow direction caused by an insufficient (leaking) valve or a communication between the left and the right heart. As such systolic, diastolic, or even continuous murmurs may be observed. Systolic ejection murmurs occur in aortic and in pulmonary stenosis (valvular or non-valvular): diastolic filling murmurs in mitral and tricuspid stenosis. Aortic and pulmonary regurgitation causes diastolic murmurs; mitral and tricuspid regurgitation causes systolic murmurs. A systolic murmur and a diastolic murmur can be observed in ventricular septal defect.

Continuous murmurs occur in patent ductus arteriosus (a connection between pulmonary artery and aorta). Musical murmurs occur as deterministic signals and are caused by harmonic vibration of structures (such as a valve leaflet, ruptured chordae tendinae, malfunctioning prosthetic valve) in the absence of flow turbulence; these are seldom observed. The location of the chest wall where a specific sound or murmur is best observed (in comparison with the other phenomena) may help in discriminating the source of the sound or the murmur [24]. These locations are dependent, not only on the distance to the source, but also on the vibration direction. Sounds or murmurs with an aortic valve origin are preferably investigated at the second intercostal space right of the sternum and those of pulmonary origin left of the sternum. The right ventricular area corresponds with the lower part of the sternum at the fourth intercostal space level, the left ventricular area between the sternum and the apex point of the heart (at the fifth intercostal space level). Furthermore, specific physiological maneuvers influencing cardiac hemodynamics may be used for obtaining better evaluation of heart sounds and murmurs. In conclusion, the existence, timing, location at the chest wall, duration, relative intensity and intensity pattern, and frequency content of murmurs and/or pathological sound complexes form the basis of auscultatory, and/or phonocardiographic diagnosis of cardiac disease.

1.11   AUSCULTATION TECHNIQUE

Before the 19th century, physicians could listen to the heart only by applying their ear directly to the chest. This immediate auscultation suffered from social and technical limitations, which resulted in its disfavor. The invention of the stethoscope and cardiac auscultation technique by Laennec (1781-1826) in 1816 (see Fig. 1.10), introduced a practical method of bedside examination, which

Figure 1.10: Rene Theophile Hyacinthe Laennec the inviter of stethoscope (Photograph courtesy of the National Library of Medicine).

became known as mediate auscultation. Over the past two centuries, many illustrious physicians have used this technique to provide an explanation of the sounds and noises heard in the normal and diseased heart. The sounds of the normal human heart can be represented by a simple onomatopoeic simulation: (. . . lubb-dup. . . .). Two sounds can clearly be identified, the first being duller than the second. A heart sound or a heart sound component is defined as a single audible event preceded and followed by a pause. As such, splitting of a sound occurs as one can clearly distinguish two components separated by a small pause. The closest splitting that can be appreciated is ≈20-30 ms. Similar guidelines are followed for the identification of phonocardiographic recordings: A sound is a complex of succeeding positive and negative deflections alternating with respect to the baseline, preceded and followed by a pause.

A sound is said to be split if a small pause between the components can be perceived. At this point, the effect of frequency filtering may be important: splitting, being invisible on a low-frequency recording, and may become recognizable on a high-frequency recording. Summarizing, in clinical PCG primarily the envelope of the recorded signal is regarded and can be stated, not the actual waveform as, for example, in ECG, blood pressure, and velocity recordings. As spectral performance of phonocardiography may exceed the possibilities of human hearing, inaudible, low-frequency phenomena can be recorded; they are also indicated as (inaudible) sounds. Acoustic phenomena originated by the heart are classified into two categories: heart sounds and heart murmurs.

Although the distinction between them is not strict, one can state that heart sounds have a more transient, musical character (cf. the touching of a string) and a short duration, whereas most murmurs have a predominantly noisy character and generally (but not always) a longer duration (e.g., a "blowing" murmur, a "rumbling" murmur). It is also believed that the genesis of both types is different: Heart sounds are indicated as types of resonant phenomena of cardiac structures and blood as a consequence of one or more sudden events in the cardiohemic system (such as valve closure), and most heart murmurs are said to be originated by blood flow turbulence. Many aspects of the problem of the genesis of these phenomena are still being discussed, including the relative importance of the valves and of the cardiohemic system in the generation of heart sounds (valvular theory versus cardiohemic theory).

Four normal heart sounds can be described in Fig. 1.12 and can be acquired from four heart sites. As shown in Fig. 1.11 they are: I, II, III, and IV (also indicated as S_1, S_2, S_3, S_4). The two having the largest intensity, that is, the first (I, S_1) and the second (II, S_2) sound, are initially related to valve closure. The third (III, S_3) and the fourth (IV, S_4) sound, appearing extremely weak and dull and observable only in a restricted group of people, are not related to valve effects. The so-called closing sounds (I and II) are not originated by the cooptation of the valve leaflets (as the slamming of a door). On the contrary, it is most probably a matter of resonant-like interaction between two cardiohemic compartments suddenly separated by an elastic interface (the closed valve leaflets) interrupting blood flow: Vibration is generated at the site of the valve with a main direction perpendicular to the valve orifice plane and dependent on the rapid development of a pressure difference over the closed valve. In the case of the first sound, this phenomenon is combined with the effect of a sudden contraction of cardiac ventricular muscle.

Pathologies of the cardiovascular system which occur due to different etiology, e.g., congenital heart valve defects, stenotic valve, and regurgitated valve as illustrated in Fig. 1.13, can affect the normal sounds with respect to intensity, frequency content, and timing of components (splitting) [25]. The first heart sound (I) occurs following the closing of the mitral valve and of the tricuspid valve, during the isovolumetric contraction period, and, furthermore, during the opening of the aortic valve and the beginning of ejection. In a medium- or high-frequency recording, a splitting of the first sound may be observed. Components related to the closing of the mitral valve (Ia, M1), the closing of the tricuspid valve (Ib, T1), and the opening of the aortic valve may be observed. There is a direct

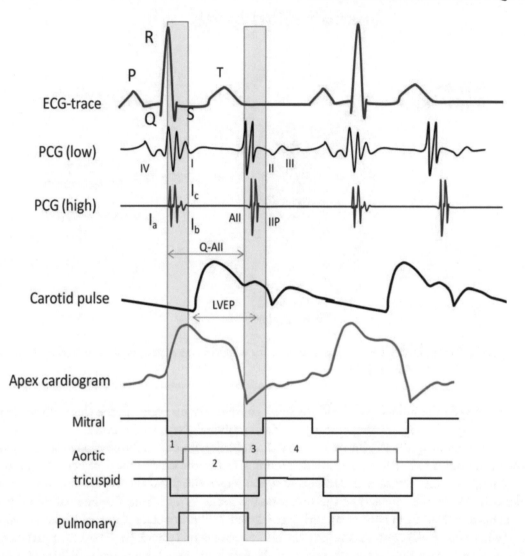

Figure 1.11: The ECG, PCG (low and high filtered), carotid pulse, apex cardiogram, and logic states (high-open) of left heart valves, mitral and aortic valve, right heart valves, and tricuspid and pulmonary valve. Left heart mechanical intervals are indicated by vertical lines: isovolumetric contraction (1), ejection (2), isovolumetric relaxation (3), and filling (4) (rapid filling, slow filling, atrial contraction). The low-frequency PCG shows the four normal heart sounds (I, II, III, and IV). In the high-frequency trace, III and IV have disappeared and splitting is visible in I [Ia and Ib (and even a small Ic due to ejection)] and in II [IIA (aortic valve) and IIP (pulmonary valve)]. Systolic intervals LVEP (on carotid curve) and Q-IIA (on ECG and PCG) are indicated.

Auscultation sites

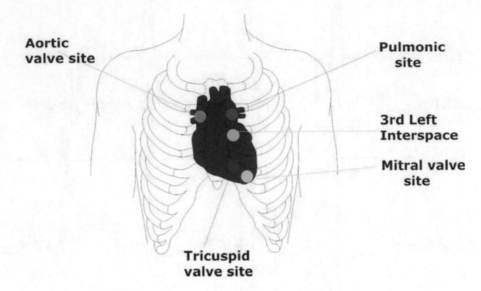

Figure 1.12: Main cardiac auscultation site used for acquiring heart sound, as indicated by American Heart Association(Ref:AHA, 1987)

relation between the intensity of I and the heart contractility, expressed in the slope of ventricular pressure rising; with high cardiac output (exercise, emotional stress, etc.) sound I is enhanced.

The duration of the PR-interval (electrical conduction time from the physiological pacemaker in the right atrium to the ventricles) is a determining factor: the shorter the time between the atrial and ventricular contraction and, consequently, the larger the distance between the mitral valve leaflets, the larger the intensity of the first sound appears. With a long PR-interval mitral valve leaflets have evolved from wide open during atrial contraction to a state of partially open to almost closed when ventricular contraction starts; the result is a weak first sound. Cardiovascular pathologies can have an effect on timing and intensities of the first heart sound components. Wide splitting is observed in right bundle branch block, tricuspid stenosis, and atrial septal defect due to a delayed tricuspid component (Ib). In left bundle branch block, Ia and Ib can coincide resulting in a single sound I. A diminished sound I is found in cases of diminished contractility (myocardial infarction, cardiomyopathy, heart failure), in left bundle branch block, mitral regurgitation, and aortic stenosis; an intensified sound I is found in mitral stenosis with mobile valve leaflets and in atrial septal defect. The second sound (II) is associated with the closure of the aortic valve and following the closure of the pulmonary valve. Splitting of the sound in an aortic (IIA, A2) and a pulmonary (IIP, P2) component is often observed. Splitting increases during inspiration as a consequence of increased difference in duration of left and right ventricular systole caused by increased right and decreased left

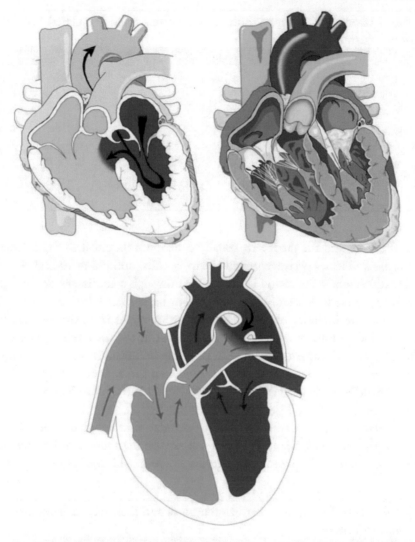

Figure 1.13: Heart valve defect is the main cause of valve stenosis and regurgitation effects.

ventricular filling; both components may fuse together at the end of expiration. Paradoxical splitting (the pulmonary component preceding the aortic one) is pathological. The pulmonary component normally has a lower intensity; an increased intensity with respect to the aortic component is generally abnormal.

As Table 1.2 indicates, the major pathological conditions of the heart valves and its correlated occurrence in the cardiac cycle heart sounds are heavily attenuated during their travel from the heart and major blood vessels, through the body tissues, to the body surface. The most compressible

Table 1.2: Major heart valves pathological conditions and its related cardiac cycle correlation occurrence

Pathology	Time	Side	Location	Position
Tricuspid stenosis (TS)	Diastolic	Parasternal	3rd ICS	Supine
Tricuspid regurgitation (TR)	Systolic	Peristernal	3rd ICS	Supine
Pulmonary stenosis(PS)	Systolic	Right	Superior	Supine
Pulmonary regurgitation(PR)	Diastolic	Right	Superior	Seated
Mitral stenosis(MS)	Diastolic	Left	Apex	Supine
Mitral regurgitation(MR)	Systolic	Left	Apex-Axilla	Supine
Aortic stenosis(AS)	Systolic	Parasternal	Superior	Supine
Aortic regurgitation(AR)	Diastolic	Parasternal	Superior	Seated

tissues, such as the lung and the fat layers, usually contribute the most to the attenuation of the transmitted sounds. To clearly perceive various heart sounds, optimal recording sites are defined, which are the locations where the sound is transmitted through solid tissues or through a minimal thickness of an inflated lung. As mentioned before, four basic chest locations exist is illustrated in Fig. 1.12 [23] where the intensity of sound from the four valves is maximized. As heart sounds and murmurs have low amplitudes, extraneous noise level in the surrounding area of the patient must be minimized. The auscultation results can be vastly improved if the room is kept as quiet as possible before auscultation begins. The patients should be recumbent and completely relaxed. They need to hold their breaths so that the noise from their breath and the baseline wandering caused by movement can be minimized [25, 27].

Table 1.3 illustrates the varieties of cardiac murmurs that are associated with systolic and diastolic phased of the cardiac cycle. It lists the focal cardiac murmurs and their related criteria of S_1 and S_2 in systolic and diastolic phases in the cardiac cycle, in addition it shows the hemodynamic

Table 1.3: List of main cardiac murmurs and their related hemodynamics criteria

cardiac murmurs	iteology (causes)	systolic criteria	diastolic criteria
MR	inefficient aortic return	$\uparrow S_1$	$\downarrow S_2$
AS	stenosis in arterial wall	$\uparrow S_1$	$\uparrow S_2$
ASD	diastolic incompetence	$\uparrow S_1$-$\uparrow S_2$	$\uparrow S_3$
TAD	abnormal conduction rhythm	$\downarrow S_1$-$\downarrow S_2$	$\uparrow S_3$
CAD	late conduction velocity	$\swarrow S_1$-S_2	$\nearrow S_2$
BAD	bradycardia initiated response	$\downarrow S_2$	$\nearrow S_1$
MI	myocardial infarction	$\uparrow S_1$	$\searrow S_2$

variations in these events. Regarding the clinical sites of which physician can acquire different murmurs as a PCG signals; these murmurs and their numerical indices of intensity and energy profile were illustrated in Table 1.4 where different clinical cases have been investigated, which represents the principal causes of cardiac murmurs.

Table 1.4: Tabulation of main heart sound auscultation site and their related S_1, S_2 intensity-energy profile

Auscultation site	S_1 intensity mV	S_2 intensity	Energy mW
p1	23.05	31.04	112
p2	26.82	35.30	123
p3	31.46	45.4	136
p4	24.20	31.7	128
p5	27.20	39.7	125

1.12 SUMMARY

The main outline of this chapter is the focus on the basic physiology of cardiac system and cardiac cycle, with intensive illustration of the heart sound associative events in this cycle. The introductory section for the main physiological basis of circulation, heart valve actions and electrical events as a bundle information was intended to give the reader an overview of how heart sounds originated. The principal mechanical and electrical events, and definitely the generation of the heart sounds in synchronization with other cardiac events (cardiac chambers blood pressure, electrical activity of the heart and respiration rate) can be synchronized of different time-traces and described in various states to represent physiological events of cardiovascular system.

CHAPTER 2

Phonocardiography Acoustics Measurement

2.1 DYNAMICS OF PHONOCARDIOGRAPHY

Mechanical heart action is accompanied by audible noise phenomena, which are easy to perceive when the ear is placed on to a person's chest wall. These cardiovascular sounds can be designated as being weak in comparison with other physiological sounds, such as speech, stomach and intestine rumbling, and even respiration noises. In fact, the latter can be heard at a certain distance from the subject, which is not true for heart noises (provided one overlooks cases of artificial heart valves). The frequency content of heart sounds is situated between 20 and 1000 Hz, the lower limit being set by the ability of human hearing. Sounds from mechanical valve prostheses may largely exceed the upper limit. Examination of cardiovascular sounds for diagnostic purposes through the human hearing sense, auscultation, has been commonly practiced for a long time [22, 25]. The only technology involved is the stethoscope, establishing a closed air compartment between a part of the person's chest surface and the physician's ear orifice.

This investigation method, however, being completely psychophysical and thus subjective, has proved its benefit and continues to be an important tool in cardiovascular diagnosis. Phonocardiography (PCG) can simply be defined as the method for obtaining recordings of cardiovascular sound, that is, the phenomena perceivable by auscultation. The origin of this method is strongly anchored in auscultation. The recordings of sounds are evaluated, on paper or computer screen, possibly in the presence of other synchronous signals (e.g., the electrocardiogram, ECG), partly psychophysically with another human sense, the eye, in examining waveform patterns and their relation with the other signals.

Phonocardiographic signals are examined with respect to the occurrence of pathological patterns, relative intensities and intensity variations, timing, and duration of events. Evidently, more objective evaluation can be performed ranging from simple accurate timing of phenomena to advanced waveform analysis and comparing recorded results with waveforms from data banks. The typical PCG signal recording was illustrated in Fig. 2.1 as it shows successive eight-phonocardiography trace with clinical annotated markers.

The importance of auscultation can be explained by the simplicity of the technique and by the strong abilities of the human ear with respect to pattern recognition in acoustic phenomena. There are different PCG signal patterns which can be differentiated from the clinical experience; some of these pattern were shown in Fig. 2.2a. For obtaining equivalent information with phonocardiography, a single recording fails to be sufficient.

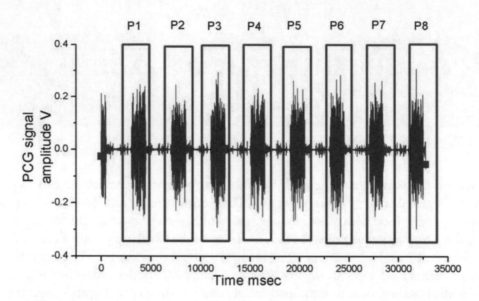

Figure 2.1: Phonocardiography trace with 8 successive S_1–S_2 waveform.

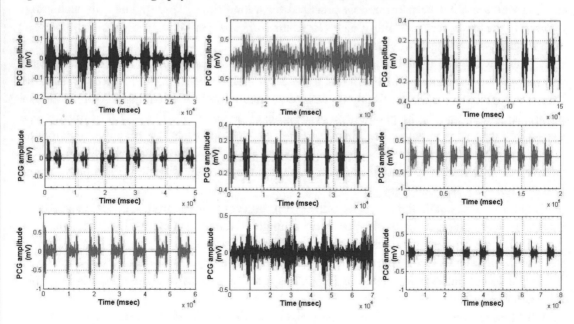

Figure 2.2: (a) PCG signal recording with different filtering coefficient for different corresponding heart sound class

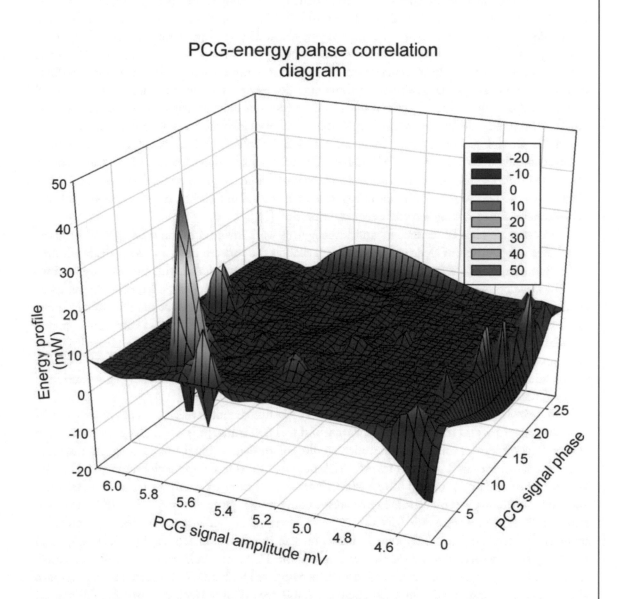

Figure 2.2: (b) Energy level of the PCG signal for 100-200Hz frequency with phase line of 0-25 rad and PCG amplifier gain range of 2 mV, observe that the maxium detectable energy occurred at low phase difference at the high PCG gain profile.

A set of frequency filtered signals, each of them emphasizing gradually higher-frequency components (by using high-pass or band-pass filters—HPF, BPF), is needed. In this way, visual inspection of sound phenomena in different frequency ranges, adapted by a compensating amplification for the intensity falloff of heart sounds toward higher frequencies, is made possible, thus rendering the method equivalent to the hearing performance: pattern recognition abilities and increasing sensitivity toward higher frequencies (within the above-mentioned frequency range). Laennec (1781-1826) was the first one who listened to the sounds of the heart, not only directly with his ear to the chest but also through his invention of the stethoscope which provided the basis of contemporary auscultation. As physiological knowledge increased through the following decades, faulty interpretations of heart sounds were progressively eliminated. The first transduction of heart sounds was made by Huerthle (1895), who connected a microphone to a prepared frog nerve-muscle tissue. Einthoven (1907) was the first to record phonocardiograms with the aid of a carbon microphone and a string galvanometer for recording muscular acoustic vibration [28].

A large number of researchers and investigators were involved in the development of filters to achieve a separation of frequency phenomena such as the vacuum tube, and thus electronic amplification became available. The evolution of PCG was strongly coupled with auscultatory findings and the development was predominantly driven by clinicians. As a result, a large variety of apparatus has been designed, mostly according to the specific needs of a clinic or the scientific interests of a medical researcher. During the 1960s, the necessity for standardization was strongly required. Standardization committees made valuable proposals [22, 27] but the impact on clinical phonocardiographic apparatus design was limited. Fig. 2.3 illustrates the PCG audible range which can be considered in synthesis and development of the accurate spectral and frequency segmentation of cardiac auscultatory waveforms.

During the 1970s and the beginning of the 1980s, fundamental research on physical aspects of recording, genesis, and transmission of heart sound was performed [25] which, together with other clinical investigations, improved the understanding of the heart sound phenomena. At the same time, ultrasonic methods for heart investigation became available and gradually improved. Doppler and echocardiography provided information closer related to heart action in terms of heart valve and wall movement, and blood velocity. Moreover, obtaining high-quality recordings of heart sound with a high signal-to-noise ratio (SNR) is difficult. Hampering elements are the inevitable presence of noise (background noise, respiration noise, muscle tremors, stomach rumbling), non-optimal recording sites, weak sounds (obese patients), and so on. Thus, interest in PCG gradually decreased. In describing the state of the art, PCG is usually compared with ECG, the electrical counterpart, also a non-invasive method. The ECG, being a simple recording of electrical potential differences, was easily standardized, thus independent of apparatus design and completely quantitative with the millivolt scale on its ordinate axis [27]. Phonocardiography has not reached the same level of standardization, remains apparatus dependent, and thus semi quantitative. Nowadays, Doppler echocardiography and cardiac imaging techniques largely exceed the possibilities of PCG and make it redundant for clinical diagnosis. The presentation of echocardiography imaging and high definition

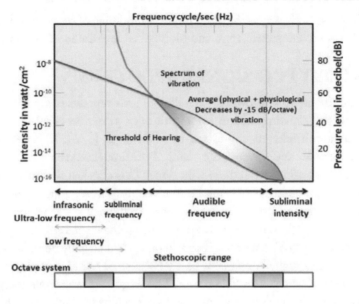

Figure 2.3: Audible range of phonocardiography signal spectrum.

of cardiac valves defect were shown in Fig. 2.4. Whereas auscultation of cardiac sounds continues to be of use in regular clinical diagnosis, PCG is now primarily used for teaching, training purposes, and research. As a diagnostic method, conventional PCG has historical value. Nevertheless, the

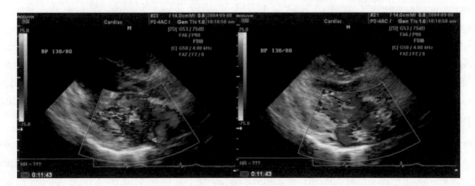

Figure 2.4: Echocardiogram of the heart valves dynamics and the blood flow dynamics through them.

electronic stethoscope (combined with PC and acquisition software), as a modern concept for PCG, may gain importance for clinical purposes. The generation of sounds is one of the many observable mechanical effects caused by heart action: contraction and relaxation of cardiac muscle, pressure rising and falling in the heart cavities, valve opening and closure, blood flowing, and discontinuation of flow. In the next section, further details will be given on the physiological significance, the physical aspects and recording methods, processing, and physical modeling of heart sounds. Special attention will also given to the electronic stethoscope and biomedical instrumentation aspects.

2.2   VIBRATORY PCG SIGNAL SPECTRUM

The displacement of the chest wall over the pericardial area represents a periodic, complex wave, which is the result of the superimposition (or addition) of pure sinusoidal waves of different frequencies and various amplitudes, as conducted by Kisie et al. [3, 6]. This definition is far more comprehensive than an older one, advocated by clinical cardiologists, which restricted phonocardiography to the recording of clinically audible vibrations. This classical definition is gradually being discarded. PCG signals can be analyzed using Fourier's analysis by separating them into a number of sinusoidal or harmonic components of the fundamental frequency $w - f$, which is that of the heart beat. It must be stressed that the complex wave representing the displacement of the chest wall constitutes a single physical entity. The method recording it should actually be called vibrocardiography but the name phonocardiography is retained, partly because of tradition and partly because sound is a physical phenomenon, whether audible or not. The frequency range and energy distribution of the audible vibrocardiography or (phonocardiography) signal were displayed in Fig. 2.5, where the threshold of the audible heart murmurs has a cut-off of 57 Hz with energy level 0.98 Dyne/cm^2, and the scheme of phonocardiography sound pressure level was shown in Fig. 2.6.

The total vibratory spectrum can be divided into various bands:

1. From 0-5 Hz. This band of vibrations corresponds to the visible and easily palpable motions of the chest wall. It includes the apex beat, the epigastric beat, and several other motions of various intercostals spaces. Studies by McKinney, Hess, Weitz, Weber and Paccon, was investigated [24, 25, 29] (cardiogram, epigastric tracing, ultra-low frequency tracing of the chest), by Johnston and Otto [23] (linear tracing), and by Edelman [24, 26](ultra-low frequency tracing, kineto cardiogram). This band is definitely subsonic because it is below the threshold of hearing. It is possible that the ballistic motions of the chest in tote are better recorded by Edelman's method (fixed pickup) while the intrinsic vibrations of the wall is better recorded by pickups which (float) with the wall itself.

2. From 5–25 Hz. This band includes the vibrations which are now called low-frequency vibrations. It barely overlaps the audible range been particularly studied by means of electromagnetic pickups which give an acceleration tracing by Rosa [6, 7, 8] and by [29, 30, 31]. It was also studied by Mannheimer [21] (displacement tracing). This band is partly infrasonic (5-15 Hz) and partly subliminal (15-25 Hz); therefore, it is partly in that range where large vibrations may be perceived by the human ear.

3. From 25-120 Hz. This band was studied by Mannheimer [29]. It was reproduced by the (stethoscopes) method of Rappaport and Sprague [30, 28], through the use of the stethocardiette (by the more modern twin beam). The most important octave band (60-120 Hz) included in this wider band is fairly well studied by the devices of Butterworth [14], Maass,

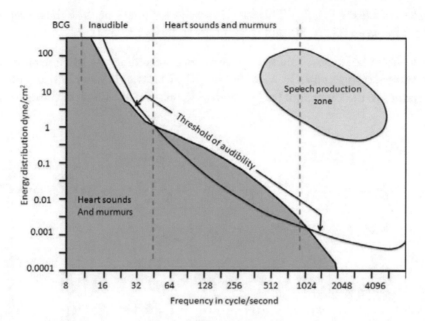

Figure 2.5: Frequency ranges and energy level of audible heart sounds and murmurs.

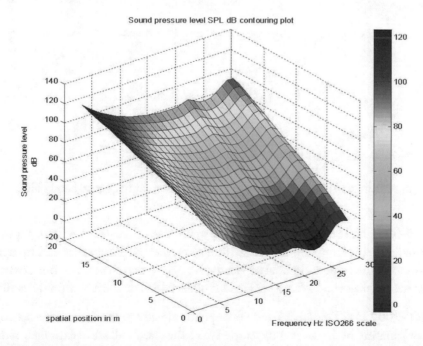

Figure 2.6: Phonocardiography sound pressure level in clinical application spectrum.

Weber, and Holldack [31, 26]. This band is partly subliminal, because the perceptivity of the aer is poor between 25 and 50, and is definitely auditory above 50.

4. From 120-240 Hz. It corresponds to the best area of recording of most apparatus and is in the auditory range. It was studied by Mannheimer[32], Maass and Weber [33] and Holldack [31]; it corresponds to the low channel of Leatham [34], one of the channels of Butterworth [33, 35].

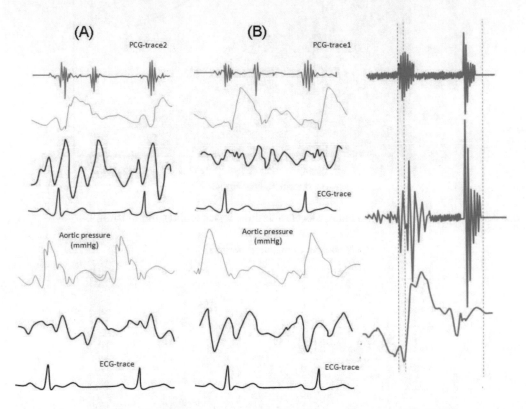

Figure 2.7: Phonocardiography signal tracing scheme for different frequency bandwidth.

5. From 240-500 Hz. It corresponds to a fairly good area of recording of many apparatus. It is approximately represented by the (logarithmic) method of Rappaport and Sprague [30]. It corresponds to the middle channel of Leatham [34], to one channel of Butterworth [34], to one channel of Maass and Weber [33], and Holldack [31]. It is still within the auditory range.

6. From 500-1000 Hz. This large band corresponds already to the area of the spectrum where sounds originating in the heart and recorded from the chest wall are of extremely reduced magnitude. Therefore, audibility may be limited or even null, not on account of frequency threshold, but on account of poor magnitude. Records have been taken in this band by Mannheimer [31],

Maass and Weber [33], Holldack [31], Leatham [34], and Luisada et al. [35]. However, most of these records are not illustrative on account of either inadequate magnification or high (noise) level. Good tracings, on the other hand, have been recorded by Luisada and Zalter through the use of a specially built phonocardiograph [35, 36].

7. From 1000-2000 Hz. This band is usually subliminal on account of poor magnitude of the vibrations. Only two apparatus seem able to record vibrations due to cardiac dynamics, in a few normal subjects and in some cardiac patients. One is that of Butterworth (full rotation) [14] and the other is that of Luisada and Zalter [35, 36]. These (high frequency) vibrations are being the object of further research in phonocardiography processing [33]. Traces for different frequency bandwidth of phonocardiography sounds were shown in Fig. 2.7.

2.3   BAND-PASS FILTER VERSUS HIGH-PASS FILTER

It has been known for a long time that (linear) cardiac transducers are inadequate for recording the vibrations of the chest wall which have the greatest clinical significance, i.e., the medium-frequency components between 100 and 300 Hz. This is because these vibrations are overshadowed by the much larger amplitude of the low-frequency components, which are simultaneously picked up by the transducer. Three methods have been developed for the study of these medium frequency components, and even more for those of high frequency:

- The use of an equalizer (microphone buffering module);

- The use of a high pass filters HPF-module;

- The use of a band pass filters LPF-module[2].

Equalizers have been preferred by certain companies because of economic considerations. Their use permits to obtain an overall picture of the vibratory spectrum. However, the use of an equalizer is equivalent to the application of a fixed high-pass filter with inadequate slope and does not lend itself to the scanning of the spectrum and study of the best frequency bands. Equalizers are being used in the apparatus of the Cardionics Company and Littmann Company.

High-pass filters have been preferred by Swedish, British, and German companies. Their use is based on the principle that, in these filters, the vibrations below the cut-off frequency are attenuated by the device while those above the cut-off frequency are attenuated by the normal (physical and physiological) slope of attenuation. Thus, a triangular curve is obtained.

A band-pass filter is a combination of a high-pass with a low-pass filter. Its use is preferred since the sharper attenuation of the high vibrations contributes to the clarity of the baseline by excluding extrinsic vibrations generated in either the room, the patient, or the apparatus.

Amplification is the degree of amplification necessary for obtaining a significant tracing obviously increases from the low bands to the high bands. Certain apparatus (like the Elema®) have a

---

[2]The filtering method of PCG signal play vital role in spectral analysis of acoustic signals.

preset degree of amplification, which automatically increases by a certain ratio when a higher band of frequency is selected. In addition, some modern E-stethoscope system like (Cardionics®) which is illustrated in Fig. 2.8, have equipped filtered the PCG trace in different frequency-bands. This ratio is based on the physical decrease of amplitude of higher frequencies, i.e., on a slope of 10 db/octave. Actually, the degree of amplification which is needed varies from case to case, and at times there is need of greater amplification for a certain band.

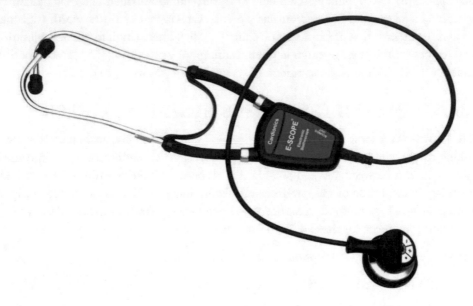

Figure 2.8: Electronic stethoscope system with ability to store and record phonocardiography traces [Ref: Cardionics Inc. 2008, USA].

In recent studies, which based on the use of a calibrated linear phonocardiograph, the use of one galvanometer only, and the observation of the degree of amplification needed in order to record the heart sounds with the same magnitude in the various bands.

This study was accomplished in several individuals, either normal or with cardiac murmurs, and in normal, large dogs. Surprisingly, it was ascertained that the amplification needed was between -4 and -8 db per octave, a range which is definitely below that of the physical decrement of vibrations (-12 db/octave so-called law of the square) and below any theoretical anticipation.

This can be explained in the following way. The heart generates vibrations of different frequencies and magnitudes. When traced phonocardiography in the various frequency bands was recorded, vibrations of the same magnitude was obtained. It is apparent that certain vibrations of medium high frequency are generated with a greater magnitude than anticipated by a purely physical law. This problem is further complicated by the transmission through the mediastinum and lungs and by the resonance of the chest wall. Further systematic and pilot-study should be performed for different phonocardiography frequency bandwidths.

2.3.1   PHONOCARDIOGRAPHY CALIBRATION

This calibration is based on the absolute linearity of the various components of a system. It was described by Mannheimer [31] and in 1939, it was revived by [31, 32]. In their apparatus, amplification is measured in decibels, and similar tracings can be taken with identical amplification for the same frequency band. At present, three methods can be used for studying selected bands of frequencies:

(a) Use of high-pass or band-pass filters for recording and comparing simultaneous (multi-channel phone) or subsequent (single channel phone) tracings of heart sounds and murmurs in various adjoining octave bands. This is the most commonly used method and was pioneered by [31].

(b) Use of a variable band-pass filter; setting of both high- and low-pass filters at the same time; and subsequent records in that extremely narrow band which is allowed to pass in such technical conditions. This method was advocated by [29].

(c) Use of a spectral analyzer, as advocated by [30, 31, 34]. This is based on a device originally developed by Fletcher for the study of speech frequencies and intensities.

Apparent splitting of 1st and 2nd heart sound in both phono-tracings was done in which stethoscopict tracing reveals a few slow vibrations in presystole; a complex first sound and a complex second sound.

In the recent medical auscultation instrumentation, the calibration procedure was made in automated steps to reduce the time and cost for a reliable and precise cardiac acoustic measurements. One of the modern automated calibrated device from (Cardionics, Inc. USA) with its stethoscope digital simulator system, as illustrated in Fig. 2.9.

2.3.2   INVESTIGATION OF NORMAL CARDIAC CYCLE IN MULTI-FREQUENCY BAND

The various phases of the cardiac cycle can be identified by means of an electrocardiogram plus simultaneous tracings of right or left atrial, right ventricular, and left ventricular pressures. This can be easily done in animals by venous and arterial catheterization. Other tracings of the vibrations of the chest wall can be recorded at the same time in the low, medium, or high frequency bands. The data obtained in this way can be compared with those obtained during clinical catheterization and with clinical tracings, where an electrocardiogram and an aortic pulse tracing is compared with phonocardiogram in various frequency bands.

The various phases of the cardiac cycle, identified by [33] and, more recently, by [34, 36], will be re-examined on the basis of the clinical data supplied by the various types of phonocardiogram modules.

Presystole. The ultra-low frequency tracing (linear) reveals a diphase wave (negative positive) at the apex and either a positive or a diphasic wave (positive-negative) at the epigastrium. According to other investigators the first phase was caused by right atrial activity and the second by left atrial activity [30]. In Fig. 2.10 (A) tracings in a normal young man of 17 was recorded, from above: Electrocardiogram, Phono (Stethoscope) at 3rd left inter-space, Phono (range 60-120) same area, Phono (range 480-1000)-same area, Plamno (range 1000-2000)-same area. Tracings 1 and 2 are

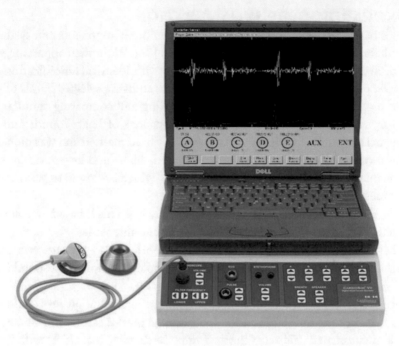

Figure 2.9: Automated phonocardiography analyser that is based on multichannel acoustic equalization. This module is also used as a training simulator for cardiac auscultation purposes. Cardionics Inc. for digital sthetoscope

simultaneous; the others have been exactly superimposed. (B) and (C) tracings recorded over the 2nd left interspace in a man with pulmonary hypertension case. The lower tracing (240/480) reveals that the pulmonary component is larger than the aortic. (C) The lower tracing (750/1000) reveals that the pulmonary component has the same amplitude as the aortic. In both cases, the atrial contraction would be transmitted, first to the respective ventricle and then to the chest or abdominal walls. The end of the second phase occurs prior to the Q wave of the ECG. The low-frequency tracing (5–25 acceleration) reveals three waves of atrial origin during the P-Q interval of the ECG according to [28], and one diphasic or triphasic wave according to [30].

The phonocardiogram in the medium frequency range (30-60 Hz) reveals a diphasic or triphasic slow wave [37]. Occasionally, three or four small vibrations can be recorded. It has been stated that this wave may fuse with the first sound and even occur alter the Q wave of the ECG [33, 37]. However, this is still open to question because the first sound itself starts with a slow vibration which is present even in cases of atrial fibrillation [36]. A possible exception is represented by children or adolescents with a short P-R interval.

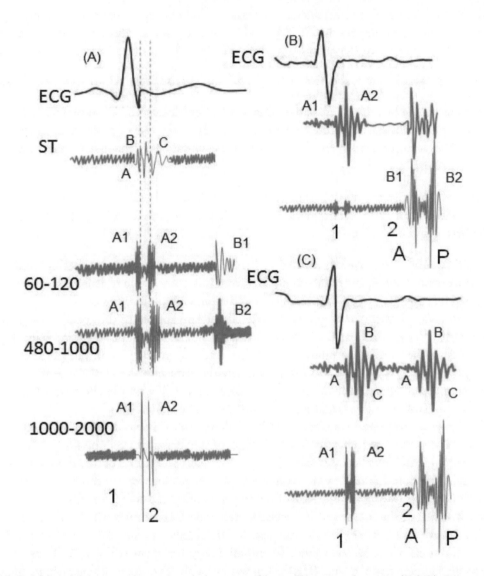

Figure 2.10: Normal young subject recorded tracing. Above: Electrocardiogram (ECG), Phonocardiography(PCG) at 3rd left inter-space, PCG (60-120Hz), PCG (480-1000Hz)-same area, PCG (1000-2000Hz). PCG tracings 1, 2, 3 are simultaneous; the others are identically superimposed. To be observed.: the division of the 1st sound to two groups of vibrations to all filtered tracings and their relationship. (B) and (C) Tracings recorded over the 2nd left inter-space in subject with pulmonary hypertension(PH). (B) The lower tracing (240/480) reveals that the pulmonary component is larger than the aortic.(C) The lower tracing (750-1000) reveals that the pulmonary component has the same amplitude as the aortic.

Ventricular Systole. The ventricular systole was divided long ago into isometric tension period (ITP) and isotonic ejection period (IEP) [34]. More recently, several further divisions were made, so that now the following research work are admitted [39, 40].

- Electro-presser latent period, from Q to the initial slow rise of (left) intra-ventricular pressure.

- Mechano-acoustic interval, from the initial rise of (left) intra-ventricular pressure(IVP) to the closure of the mitral valve and the beginning of the rapid rise of pressure in the ventricles. This phase is terminated by the first group of rapid vibrations of the first sound and was called entrant phase by [37].

- Phase of rapid rise of pressure, from the closure of the mitral valve to the opening of the aortic valve and, in the phonocardiogram, from the first to the second main group of vibrations of the first sound S−1 [43, 45].

The phase of expulsion or ejection was divided by [37, 38] into maximal ejection, lasting until the top of the aortic pressure curve, and reduced ejection, from this point to the incisura of the aortic pulse; it is followed by relaxation of the ventricles during the phase of protodiastole.

The ultra-low frequency tracing (linear) with the pickup applied to the apex [44, 47]; Kineto-cardiogram [49] often reveals two distinct waves, one during the entrant phase (or mechano-acoustic interval), and another during the phase of rapid rise (actual isometric phase). The phase of ejection is typically revealed by a (systolic collapse), caused by the reduced volume of the ventricular mass, unless motion of the apex maintains contact of the cardiac wall with the chest wall and causes a systolic plateau. End of systole is marked by a positive wave or notch [47].

The low-frequency acceleration tracings [42, 43] show a complex ABC during the RS complex of the ECG. This coincides with the first slow vibration of the first sound and the slow rise of pressure in the ventricles (mechano-acoustic interval). A second complex CDE occurs during the S wave and the ST junction; it coincides with the first group of large vibrations of the first sound. A third complex EFG occurs during the upstroke of the carotid tracing and coincides with the second main group of vibrations of the first sound. Afterward, there is a GHI complex which coincides with the rise of the T-wave and the carotid shoulder, and an IJK complex at the peal of T, which ends with the initial vibrations of the second sound. To amplify the gain response of this PCG low frequency band by using digital infinite impulse (IIR) low-pass filter with 6th order to attenuate high-frequency component in PCG trace. Fig. 2.11 represents the IIR-LPF (low-pass filtering) response of two PCG signal trace. This response indicating the robustness of combining LPF-filtering for synchronous acquisition of two PCG trace.

The phonocardiogram in the medium-low range (60-120 Hz) shows a small initial vibration of extremely low frequency and magnitude during the mechano-acoustic interval. It then shows a central phase of four large diphasic vibrations (Luisada, [35]), which can often be divided in two main groups coinciding with the valvular events of the heart [44, 45]; these are separated in the adult by an interval of 0.04-0.05 sec. It has been shown that the peaks of these large vibrations may be

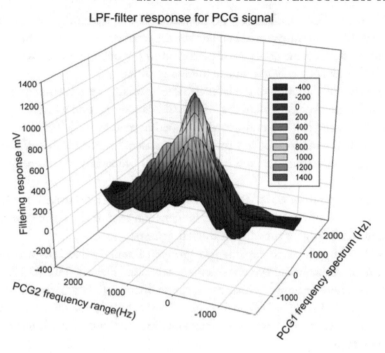

Figure 2.11: Filter response of low-pass filter apply to phonocardiography signal. The response vary in a nonlinear zone due to redundant noise in the stethoscope unit.

identified through right and left heart catheterization, and can be shown to coincide with the four valvular events [44].

These, as known, succeed each other in the following sequence: Mitral closure, tricuspid closure, pulmonary opening, aortic opening [36]. For this reason, the symbols M, T, P, A, referring to the four valves, were suggested for the main four vibrations, if they can be identified [43].

It is unfortunate that not in all persons such a clear cut distinction can be made and that additional, smaller waves damaged at times this desirable clear cut picture. It is obvious that, if both valvular closures and valvular openings are accompanied by vibrations, the interpretation of the mechanism of production of the first heart tone (or sound) will require a revision. Subsequent to the opening of the aortic valve, the medium-low frequency tracing often presents from one to three vibrations in decrescendo which seem connected with the early phase of ejection and which usually terminate with the peak of the aortic pulse. They have been explained with vibrations of the aortic and pulmonary walls. The second half of systole is usually clear of vibrations, but there may be one or two small vibrations during mid-systole.

Medium-high frequency vibrations bands [120-240 Hz and 240-480 Hz]. In these bands (or only in the latter), a fairly good unitary pattern occurs, as shown by [44]. The first heart tone often becomes split into two phases, separated by an interval of 0.04-0.05 sec; or there are two larger vibrations within a series of 3–5 smaller ones. There is no discussion about explaining the first larger vibration of normal subjects with the mitral closure, in regard to the second, which Leatham [34, 35] explained as due to tricuspid closure; on the contrary, explanation of the opening of the aortic valve [33] was performed. This interpretation is based on the following facts:

(1) The initiation of activation of the left ventricle normally precedes that of the right by a very small interval. A mechanical precession of the left ventricular contraction was already proved in dogs, by Hamilton and confirmed by [36], and was ascertained in man by [38]. This precession is only 13 msec, and has individual variations of not more than 4 msec. In other words, the synchronization between the starting of contraction of the two ventricles is practically not more than 17 msec and may be only 9 msec.

There may be respiratory variations; this interval is further decreased by the fact that the methane-acoustic interval of the left ventricle is longer than that of the right: closure of the mitral valve is slightly delayed by higher pressure in the left atrium.

Therefore, it is impossible to explain two groups of sound vibrations usually separated by 0.04-0.05 sec. with two mechanical events which are separated by only 0.01-0.015.

(2) The interval between mitral closure and aortic opening (left ventricular isometric tension period) was evaluated in the dog by Whoor's as of the order of 0.05 sec interval. and Braunwald et al. for man with a 0.06 sec interval, which is practically identical with that found between the two above groups of vibrations.

(3) The interval between the first and the second large vibration does not vary when right ventricular contraction is experimentally delayed, and may even increase when the latter is not delayed. It is obvious that a delay of right ventricular contraction should increase the interval between the large vibrations if the second group of vibrations were due to tricuspid closure.

(4) In cases of mitral stenosis and atrial fibrillation, it is accepted that mitral closure is delayed and is either simultaneous with or follows tricuspid closure. Short diastole would theoretically further delays mitral closure on account of higher left atrial pressure. However, in these circumstances, one may find a split first sound with an interval of 0.05-0.06 sec between the signal components.

If the second was (mitral), the interval of (0.15 sec) between Q of the ECG and this sound, which was found, would be far too long to be accepted.

(5) The second large vibration coincides with the rise in pressure of the aorta and may coincide with or slightly below that of the pulmonary artery. The precession of this sound over aortic rise of pressure, may have been due to incorrect interpretation of that small rise of pressure in the aorta, which occurs during isometric contraction [39].

This leads to discussion of the so-called ejection sound (so named by Leathiam). Such a sound is a new phenomenon which arises in cases of stenosis of the aorta or phonic valve or in cases with dilatation of the aorta or pulmonary artery and allows the opening of the semilunar valves.

On the contrary, McMontry since 1953 [33] concluded that such a sound represents, in the majority of cases, an accentuation and delay of the second group of vibrations of the first sound, possibly related to an abnormal gradient of pressure across the valve or disturbed flow in the vessel.

(In particular, it represents the accentuation of either (P) or (A) according to whether the pulmonary valve or artery is involved, or the aortic valve or aorta.) Lie et al. [39] also admitted that, in certain cases with dilatation of one of the large arteries, a slower vibration occurs during the early phase of ejection (ascending 1) ranch of the pulse). This contention is proved by the following facts:

1. It is not clear in previous research the ability to demonstrate three groups of high-pitched vibrations (mitral closure, tricuspid closure, ejection sound) over the same part.

2. Catheterization of either pulmonary artery or the aorta shows the rise of pressure coinciding with and not following the so-called ejection sound [41]. Studying this large sound in cases of congenital heart disease confirmed that it represents a pathological accentuation (and occasionally a delay) of a normal component of the first tone.

2.3.3   HIGH-FREQUENCY VIBRATIONS BANDS [500-1000 HZ AND 1000-2000 HZ]

These can be recorded only through great amplification (60–100 dB) and only in young individuals with a thin or flat thorax wall. The highest band, particularly, is only exceptionally recorded. On the other hand, an intermediate band (750-1500) can be studied in a greater number of individuals. At or within the apex, and sometimes also in the 3rd left interspace, one can obtain either one or two large vibrations. When only one is recorded, it coincides with the first larger vibration of the medium-high bands; when two are recorded, the second coincides with the second larger of such bands. Occasionally, either of the larger vibrations is accompanied by a smaller one, either before or after onset.

As in the previously described bands, the investigations explain these vibrations of high frequency and low magnitude as coinciding with the events of the left heart (mitral closure-aortic opening). It is interesting to note that occasionally a tiny vibration at mid-systole coincides with the G peak of the low-frequency tracing or occurs in late systole. It is a general rule that the vibrations of high frequency and small intensity (300 Hz and above) take place at the beginning of each sound, tone, or component. They are immediately followed by vibrations which have a lower frequency, a greater intensity, and often a longer duration.

If the tracing is recorded at the base, usually the aortic component is revealed by a large vibration while the pulmonary is revealed by either a tiny vibration or not at all. In severe pulmonary hypertension, the pulmonary component is often larger than the aortic and may even be the only one recorded. At the midprecordium or apex usually one can record either only the vibration corresponding to mitral closure (1st tone) or this plus that of aortic closure (2−nd tone).

Diastole phase according to [37], can be divided into rapid filling and mid slow filling (or diastasis). It is interesting that newer studies [40, 41] have shown that the phase of rapid filling is partly aided by the elastic recoil of the ventricular walls causing a partly active diastole.

Proto-diastole Isometric Relaxation. The former phase, according to [42], lasts from the beginning to the peak of the incisura of the aortic tracing (closure of the aortic valve); the latter, from the closure of the aortic valve to the opening of the mitral valve.

2.3.4 ULTRA-LOW FREQUENCY TRACING (LINEAR FREQUENCY BAND)

This tracing normally presents a descending limb to the peak IIa (closure of the aortic valve) to the trough O (or IIb) which marks the lowest point of the tracing [43]. This point indicates tricuspid opening, if the tracing is recorded at the epigastrium (often, in this area, there is a mirror-like pattern: low point II a, peak IIb), and mitral opening, if it is recorded at the apex [41, 43]. In low-frequency tracing (5-25 acceleration), the end of the T-wave of the ECG and the first vibration of the 2nd heart sound are simultaneous with the K-wave of this acceleration tracing. Closure of the aortic valve is accompanied by the complex KLM, while the peak M coincides with the opening of the AV valves [44].

During this phase, the tracing rises from the lowest point (point (O) or IIb) to a high position, marking the maximum of rapid filling and coinciding with the 3rd sound. This part of the tracing is the most commonly reproducible and, therefore, the most useful for identifying a 3rd sound [39].

The peak usually coincides with the 3rd heart sound and the maximal of the phase of rapid filling. The OP complex occurs during the phase of diastasis, if this phase is not abbreviated by tachycardia.

2.3.5 MEDIUM-LOW FREQUENCY TRACING [60-120 HZ]

This tracing usually reveals the complex of the second heart tone with two-to-four large vibrations. Frequently, two larger components can be recognized within the central part of this tone (aortic and pulmonary closures) [34, 42]. The opening of the mitral valve is not revealed by this tracing in normal subjects. Pavlopoulos et al thought that, in race normal cases, a small vibration of lower frequency occurred at the time of mitral opening [35]. However, in retrospect, the vibration might have been a deformed picture of the pulmonary component, even though recorded at the 4th left interspace. On the other hand, in cases of mitral stenosis, this vibration is well recorded (opening snap).

In this phase, a small vibration may occur in coincidence with tile peak of rapid filling (3rd sound). It may be much larger in cases with increased pressure of the atria (triple rhythms or gallops) and it may be split [45], thus simulating the occurrence of a 5th sound [48].

2.3.6 MEDIUM-HIGH FREQUENCY BAND [120-240 HZ AND 240-480 HZ]

In these bands, two large vibrations emerge within the 2nd heart tone. There is no discussion among the various researchers in the identification of the first with the closure of the aortic valve, and the second with that of the pulmonary valve. The best place for recording both of them is the 3rd left interspace. Usually only the first of them (aortic component) is transmitted to the 2nd right interspace and toward the apex while the second (pulmonary component) may do so in cases of pulmonary hypertension.

A delay in the pulmonary component occurs in pulmonary hypertension, pulmonary stenosis, and right bundle branch block. A delay in the aortic component, on the other hand, occurs in aortic stenos is or left bundle branch block. The mitral opening snap is recorded best in these bands. It is often of small amplitude and high pitch.

In normal subjects, no vibrations are recorded. In cases with pathological triple rhythms, the 3rd sound is usually best recorded in the band 120-240 but, in certain cases, is well recorded above 500 and may reach even higher frequencies. This points out the difficulty of an absolute differentiation between a (gallop) sound and an (opening snap) of the mitral valve on the basis of frequency alone.

The opening snap, usually of a higher pitch than the (gallop), is best shown in the bands 120-240 or 240-480 (occasionally even higher), and is still recorded as a small twitch above 750 and even above 1000. Therefore, even though the opening snap is higher pitched than the (gallop) sound, there is an overlapping of the bands in which they are recorded best, and other elements have to be taken into consideration for the differential diagnosis.

2.3.7   THE HEART TONES PRODUCTION MECHANISM

Numerous studies have been devoted to this problem, and in particular to the mechanism of the first heart tone. We shall quote here only those from our group [31, 33, 36], which discuss previous experimental work by others and contribute to clarification of this problem. According to recent interpretations [29, 31], the vibrations of the heart tones are due to the rapid changes in the pressure of the blood (and of the cardiac walls surrounding it) whenever a rapid acceleration or deceleration occurs in coincidence with, but not caused by, the valvular movements.

Aortic valve opening, for example, will cause a sudden acceleration of the left ventricular blood, which is being ejected, plus a ram effect of this blood against the static blood contained in the aorta. In an elastic chamber filled with fluid, any sudden motion throws the entire system into vibration, the momentum of the fluid causing an overstretch of the elastic walls, followed by a recoil and a displacement of fluid in the opposite direction. The intensity of a sound seems to be proportional to the rate of change of the velocity of the blood while its frequency seems to be connected with the relationship between vibrating mass and elasticity of the walls. Studies of the heart tones made in the different frequency bands have shown [38, 43] that all four valvular events involved in the first tone are revealed by a medium-low frequency band (60-120 Hz).

On the contrary, the higher bands, and especially that between 500 and 1000, usually reveal the events of the left heart: a vibration due to mitral closure is best recorded in the 3rd-4th interspaced (1st tone); a vibration due to aortic closure is best recorded at the base (2nd tone). Pulmonary closure (2nd tone) is revealed in this band as a small vibration, except in cases of severe pulmonary hypertension.

The 3rd and 4th tones are usually best revealed by the ultra-low and low-frequency bands (apex cardiogram or (kinetocardiogram); low-frequency acceleration tracing in the 5-25 band; displacement tracing in the 15-30 band). It is interesting to note that the high-frequency tracing may occasionally reveal a tiny vibration in systole which corresponds to the H wave of the low-frequency tracing. This indicates the existence of a small, high-pitched overtone coincides with the peak of the aortic pulse.

The 3rd and 4th tone have been the object of numerous studies. The opinion which seems to gather the best experimental and clinical support is that they are caused by the onrush of blood into the ventricles during the two phases of accelerated filling in early diastole and presystole. As such, they seem to be generated in the ventricular walls, and intracardiac phonocardiogram confirms this point of view. Earlier vibrations recorded by the esophageal method may be more closely connected with the atrial contraction (4th tone). Speculations dealing with a valvular origin of these tones are periodically presented, in regard to the 3rd tone, they do not seem to be acceptable. In regard to the 4th [37], it is likely that two separate components, one valvular and one myocardial, occasionally take place. It is still open for discussion, however, whether the valvular component is recorded in cases other than complete AV-block pathology.

2.4   STETHOSCOPE TRANSDUCER MODELING

2.4.1   MICROPHONE TRANSDUCER

There are different types of microphones are suitable for picking up air-coupled body sounds from the skin. These include the following entitites:

- Capacitor microphones, where the induced vibration of a metalized mylar film (forming one plate of a capacitor) changes the capacitance between it and a fixed plate, inducing a change in the capacitor voltage under conditions of constant charge.

- Crystal or piezoelectric microphones, in which air-coupled sound pressure vibrates a piezo-crystal, directly generating a voltage proportional to (dp/dt), where p is the sound pressure at the microphone.

- Electret microphones are variable capacitor sensors in which one plate has a permanent electrostatic charge on it, while the moving plate varies the capacitance, inducing a voltage which is amplified. Electret microphones are small in size, and found in hearing aids, tape recorders, computers, etc.

Microphones generally have a high-frequency response that is quite adequate for endogenous body sounds. It is their low-frequency response that can be lacking. Indeed, some heart sounds are subsonic, ranging from 0.1–20 Hz (Webster, 1992), while 0.1–10 Hz is generally inaudible, and sound with energy from 10 to 20 Hz can be sensed as subsonic pressure by some listeners.

To record body sounds, a pair of B&K model 4117 piezoelectric microphones was modified to cover down to < 1 Hz by inserting a fine, stainless steel wire into the pressure relief hole that vents the space in back of the piezo-bender element. The wire increased the acoustic impedance of the vent hole and thus increased the low-frequency time constant τ of the microphone from about 0.05 sec (corresponding to a −3dB frequency of c.a. 3 Hz) to < 0.15 seconds, giving a −3dB frequency < 1 Hz.

The high (−3dB) frequency of the 4117 microphones was 10 kHz. The voltage sensitivity of the M4117 microphone at mid-frequencies range is about 3 mV/Pa (3 mV/10 mbar). Typical cardiac

3D microphone internal structure was shown as in Fig. 2.12, where it consist of four main components; (1) external crystal (piezoelectric); (2) ground-cage, (3) oscillating drum, and (4) excitation source.

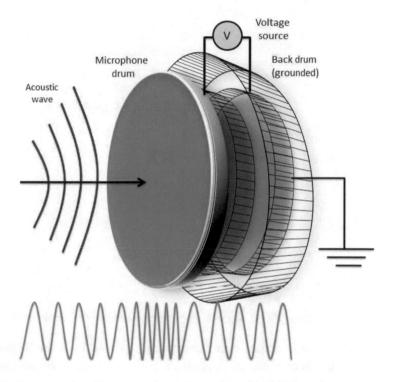

Figure 2.12: 3D diagram of cardiac microphone structure and equivalent electrical circuit.

Another high-quality B&K microphone of the model 4135 quarter-inch condenser microphone was used. This research-grade device had a high-frequency, 3dB frequency in excess of 100 kHz and a total capacitance of 6.4 pF with a diaphragm-to-plate spacing of 18 μm. For body sounds, the low-frequency end of the 4135's frequency response is of interest. Three factors affect the 4135 microphone's frequency response:

1. The acoustic time constant formed by the acoustic capacitance (due to the volume between the moving (front) diaphragm and the insulator supporting the fixed plate), and the acoustic resistance of the small pressure equalization tube venting this volume. As in the case described before, the acoustic resistance can be increased by inserting a fine wire into the tube; this raises the acoustic time constant, and lowers the low −3dB frequency.

2. The low −3dB frequency is affected by the electrical time constant of the parallel RC circuit shunting the microphone capacitance as illustrated in Fig. 2.13.

3. The mechanical resonance frequency of the vibrating membrane and its mass generally set the high-frequency end of the microphone's response. The smaller and thinner the diaphragm, the higher will be its upper -3dB frequency.

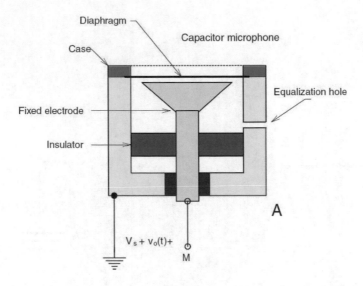

Figure 2.13: A cross-section of a capacitor microphone used in phonocardiography acquisition.

The capacitance change of a microphone over time can be expressed as in the following equation, where C_0: the output capacitance of the microphone, δC: change of microphone capacitance, and ω: microphone cage natural frequency of oscillation.

$$C(t) = C_0 + \delta C \sin(\omega t). \qquad (2.1)$$

This expression for C(t) is substituted in the voltage equivalent microphone equation, and the resulting equation is differentiated with respect to time. This results in a first-order nonlinear ordinary differential equation (ODE) in the loop current, i(t), which is solved to yield a frequency response function, which can be written as follows:

$$\begin{aligned} i(t) = \quad & \frac{V_s \delta C / C_0}{\sqrt{[R_s^2 + 1/(\omega C_0)]}} \sin(\omega t + \phi_1) \\ & - \frac{V_s R_s \delta C / C_0^2}{\sqrt{[R_s^2 + 1/(\omega C_0)]}} \sin(2\omega t + \phi_1 + \phi_2) \\ & + \text{ Higher-order harmonics,} \end{aligned} \qquad (2.2)$$

where V_s is the microphone DC excitation voltage and R_s is the source impedance. Note that $\phi_1 = tan^{-1}[1/(\omega R_s C_0]$ and $\phi_2 = tan^{-1}[1/(\omega 2.R_s C_0]$. When $\delta C_0/C << 1$, the fundamental frequency term dominates, and the ac small-signal output of the microphone (superimposed on the dc voltage,

V_s, can be written as a frequency response function:

$$\frac{V_*}{\delta C}(j\omega) = \frac{V_s R_s \omega}{\sqrt{1 + (\omega C_0 R_s)^2}}.$$ (2.3)

Olson et al. [15] points out that this is the same result obtained if placing an open-circuit (Thevenin)

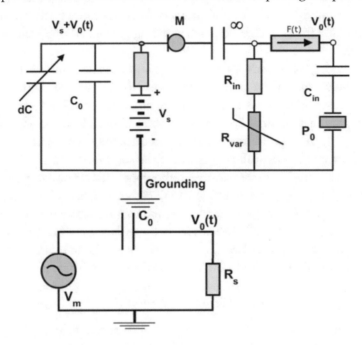

Figure 2.14: (a) Schematic diagram of equivalent circuits which represents the cardiac microphone system; (b)r Simplified linear circuit of the cardiac capacitive-microphone.

voltage of $V_{oc} = V_s(\delta/C_0 sin(\omega t + \phi_1))$ in series with C_0 and R_s in the loop, and observe $v_o(t)$ across R_s. From Equ. 2.3, we see that the low corner frequency is $f_{Lo}=1/(2\pi R_s C_0)$Hz. For example, if C_0=7 pF, and R_s=10 ohms, then f_{Lo}=2.3 Hz. The equivalent electrical circuit of the cardiac microphone is illustrated in Fig. 2.14.

2.4.2 ACOUSTIC COUPLING

No matter which kind of sensor is used to detect endogenous sounds waveform over the patient skin, there is a problem in efficiently coupling the sound mechanical vibrations from within the body to the microphone (or the eardrum). Since the days of Laennec and his first stethoscope, a bell-shaped or conical interface has been used to effectively couple a relatively large area of low-amplitude acoustic vibrations on the skin to a small area of larger amplitude vibrations in the ear tube(s).

This bell-shaped interface is in fact an inverse horn (at that time, literally cow horns), which were used pre 20th century as hearing aids. Note that the pinna of the human ear is an effective inverse horn, matching the low acoustical impedance of open space to the higher impedance of the ear canal and eardrum. Like all horns, the pinna exhibits combfilter properties, attenuating certain high frequencies in narrow bands around 8 and 13 kHz [37]. Regular horns were first used as speaking trumpets, later as output devices for early mechanical record-players; here the lateral displacement of the needle on the disk vibrated a mica diaphragm (c. 2 in. in diameter). A horn was used to couple those vibrations to a room and listeners. Most acoustics textbooks describe horns in the role of coupling sound from a small-diameter, vibrating piston to a large-diameter opening into free space (the room).

In examining endogenous body sounds, the opposite events occur. A large area of small-amplitude acoustic vibrations on the skin is transformed by the inverse horn to a small area of large-amplitude vibrations (at the eardrum or microphone). It is beyond the scope of this section to mathematically analyze the acoustics of direct and inverse horns. However, we will examine them heuristically. Basically, horns and inverse horns are acoustic impedance-matching systems. They attempt to couple the acoustic radiation impedance of the source to the acoustic impedance of the horn termination. The termination in the case of a stethoscope is the rather complex input impedance of the coupling tubes (or tube, in Laennec's instrument); in the case of a microphone, it is the moving diaphragm. If impedances are not matched, sound transmission will not be efficient, because there will be reflections at interfaces between any two media with different acoustic impedance characteristic; e.g., at the skin–air interface, and at the air–microphone interface. The characteristic acoustic impedance of a medium is a real number defined simply by [37]:

$$Z_{ch} = \rho(c)cgs.ohms.(ML^2T^1). \tag{2.4}$$

For air, the density, ρ, is a function of atmospheric pressure, temperature, and relative humidity. The velocity of sound, c, in air is not only a function of atmospheric pressure, temperature, and relative humidity, but also of frequency. Thus, the Z_{ch} of air can vary over a broad range, varying from c. 40–48 cgs ohms (43 is often taken as a compromise or typical value). An average Z_{ch} for body tissues (skin, muscle, fat, connective tissue, organs, blood) is c. $2\text{x}10^5$ cgs ohms. Thus, we see that there is an enormous impedance mismatch in sound going from the body to air, and much intensity is reflected back internally. Clearly, there will be better sound transmission through the skin when the skin sees a much larger acoustical impedance looking into the throat of the inverse horn.

2.5   SUMMARY

The summary of this chapter is as follows: Different PCG signal bandwidth was illustrated and the PCG filtering concepts, calibration, and system approach have been considered. Cardiac microphone and stethoscope system with PCG-signal acquisition calibration were explained in response to active-input signal scheme. The modeling of cardiac microphone as equivalent circuit was also demonstrated with a listing of main types of microphone and it's application in auscultation strategy. The impedance circuit for the acoustic signal propagation inside thorax cavity.

CHAPTER 3

PCG Signal Processing Framework

3.1 PHONOCARDIOGRAPHY SIGNAL PRESENTATION

Phonocardiography (PCG) is a useful technique for registering heart sounds and murmurs. It was developed in the early days to overcome the deficiencies of the acoustical stethoscope properties. A phonocardiogram, which is the graphic representation of the heart sounds and murmurs, can document the timings and annotates the different relative intensities of heart sounds and murmurs.

Cardiologists can then evaluate a phonocardiogram based on the changes in the wave shape i.e., morphology) and the timing parameters (temporal characteristics). A number of other heart-related variables may also be recorded simultaneously with the phonocardiogram, including the ECG, carotid arterial pulse, jugular venous pulse, and apex cardiogram. Such information allows clinicians to evaluate the heart sounds of a patient with respect to the electrical and mechanical events of the cardiac cycle.

Although phonocardiography can record and store auscultatory findings accurately, its usage as a diagnostic tool is quite uncommon because of the required critical procedures and complicated instrumentation. A standard procedure to record the phonocardiograms requires a specially designed, acoustically quiet room [41, 42, 43]. In terms of equipment, because phonocardiographic devices were introduced before compact analog integrated circuits were available, they are typically large, noisy, and inconvenient to use. With the introduction of the electronic stethoscope, phonocardiography may possibly make a comeback in clinical practice.

The newly developed electronic stethoscopes are more compact, robust to noise, immune to disturbance, and much more convenient for diagnostic use. They permit a digital recording of heart sounds, and therefore allow clinicians to perform analysis of the phonocardiogram on a PC platform. The phonocardiogram may have even greater diagnostic importance in the future as further improvements are made in the electronic measurement technology and signal processing algorithms.

These algorithms can help to represent the heart sound as a set of characteristic signal components, which may form the detection basis of various heart diseases. Much of the research effort has been devoted to the exploration of signal processing methods with reduced sensitivity to the recorded noise and improved identification of the exact start and end points of major heart sounds (i.e., S_1, S_2, and murmurs). In general, three ways to represent and characterize heart sound exist:

- spectrum estimation;

- time-frequency analysis;

- nonlinear analysis.

The most popular method to represent heart sound components in the past was spectrum estimation. A variety of techniques have been developed to analyze the frequency contents of heart sound. Earlier studies used spectral estimation techniques to extract features in the frequency domain [44, 45]. Classical methods for power spectral density (PSD) estimation employ various windowing and averaging functions to improve the statistical stability of the spectrum obtained by Fast Fourier Transform (FFT) (.g., the Welch periodogram method).

Early researches have demonstrated some success in distinguishing normal from abnormal patients based on the average power spectrum of diastolic heart sounds, estimated by traditional FFT methods [45, 46]. Despite the success, classical methods may not provide an accurate power spectrum when the signal-to-noise ratio (SNR) is low and the length of the desired signal is short [45]. An alternative spectral estimation technique is the parametric modeling method (e.g., autoregressive (AR), moving average (MA), and autoregressive moving average (ARMA)). Parametric modeling involves choosing an appropriate model for the signal and estimating the model parameters. These parameters can then be used to characterize the power spectrum of the signal, to classify the signal, or to perform data compression and pattern recognition tasks.

Among these modeling techniques, AR modeling shows outstanding performance when the signal has very sharp peaks. The ARMA model can be used to model signals with sharp frequency peaks and valleys. The studies have shown that the application of parametric modeling methods to signal identification can provide a good estimation of spectral features, particularly for a signal with low SNR, which led to the use of model-based methods in the analysis of heart sounds and the detection of features associated with coronary artery diseases [50, 52].

Another class of spectral estimation techniques is nonparametric Eigen-analysis based method, which is based on Eigen decomposition of the data or its autocorrelation matrix. The eigenvector method (or the minimum-norm method) can be used to extract the signal buried in noise. These methods should be extended in the next chapters with a deepening in spectral estimation approach.

In theory, the eigenvector method has infinite resolution and provides an accurate spectral estimation regardless of the SNR. In relevant studies of acoustical detection of coronary artery disease by Akay et al., the eigenvector method has demonstrated the best diagnostic performance when compared with the other spectral estimation methods like FFT, AR, and ARMA [51]. In general, the parametric spectral estimators (AR and ARMA) and the eigenvector methods offer the promise of higher resolution over the FFT. However, the major shortcomings of AR, ARMA, and eigenvector methods are that in each case poor spectral estimation occurs if the assumed model is inappropriate or if the model orders chosen are incorrect [50].

Early methods of frequency domain analysis were mostly based on the direct applications of Fourier transform or autoregressive spectral estimation techniques. However, if the statistical properties of the signal change with time (i.e., non stationary), the direct application of these techniques may be inappropriate as important time events like frequency variation would be lost in the transformation process. As heart sounds are non stationary in nature, more recent research adopts the time-frequency analysis in order to capture the temporal variation in the heart sound signal [56, 57].

3.2   DENOISING AND SIGNAL FILTERING TECHNIQUES

One of the major problems with recording heart sounds is noise parasitic effects. In practice, the noise source may be composed of instrumentation noise, ambient noise, respiratory noise, thoracic muscular noise, peristaltic intestine noise, and fetal breath sounds if the subject is pregnant.

The contribution of each source may vary significantly depending on the technical characteristics of the recording instrumentation, sensor detection bandwidth, the recording environment, and the physiological status of the subject. Currently, no way exists of knowing a priori what the particular noise component is, or determining the noise component once measurements have been made.

A reasonable solution to noise reduction can be carried out in two parts. First, extraneous noises must be minimized in the vicinity of the patient during recording. Second, various signal processing methods, such as notch filtering [54], averaging [57], adaptive filtering [58, 59], and wavelet decomposition [58, 61], can be designed and implemented by hardware or software to remove the noise, based on the assumption that the noise is an additive white noise. Although these methods have been prove effective and robust results, more research is required to determine what kind of noise and disturbance could corrupt the recorded heart sounds so that a system could employ different denoising (filtering) techniques based on the specific noise present.

3.3   PCG SIGNAL PRESENTATION

Physical modeling aims at the localization of a specific sound source in the heart and, by analyzing the externally recorded vibration signals, at the quantification of the constitutive properties of the cardiac structures involved (e.g., stiffness of a valve leaflet, myocardial contractility) and of the driving forces, which set these structures into vibration.

The physical situation is extremely complicated. The vibration source is situated within the cardiac structures (having viscoelastic properties) containing and driving blood (a viscous fluid). The transmission medium, (the tissues between the heart and the chest wall) is viscoelastic and inhomogeneous. Transmission in such a viscoelastic medium implies compression and shear waves, which both contribute to the vibrations at the chest wall [60]. It is not simply a problem of acoustic pressure as in a perfect fluid. Distortion due to transmission seems obvious.

In order to study transmission and to relate chest wall vibrations to properties of cardiac structures and hemodynamic variables, advanced signal processing techniques are used. A broad review is given by Durand et al. [57, 60]. As the chest wall vibratory phenomenon is represented by a spatiotemporal kinematic function, it can principally be approached in two ways: by sampling in time, as a set of images of chest wall movement, or by sampling in space by a set of time signals obtained with multisite recording. Multi-site heart sound recording implies a large set of pickups

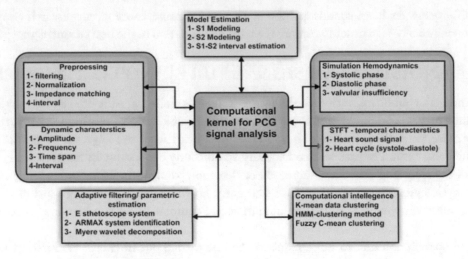

Figure 3.1: Phonocardiography signal processing loop.

(preferably light-weight, thus inducing minimal loading). In this way, spatial distribution of vibration waveforms on the chest wall can be derived. Based on the results of such a method a physical model for heart sound genesis has been presented that can analytically be solved in a viscoelastic medium: a sphere vibrating along the axis of the valve orifice [60].

This mechanical dipole model agrees to the idea of sound generation as a resonant like vibration of the closed elastic valve leaflets and the surrounding blood mass. With this model a typical inversion of vibration waveforms on the chest wall could be explained: The phase reversal is most expressed for the second sound, according to the anatomical position and direction of the aortic orifice. The model has been used to calculate source functions (the inverse problem). Spatial parameters on vibration waveforms have been formulated (22-25 mV) [61].

The phonocardiography signal processing loop was shown in Fig. 1.3, in which inside the PCG kernel signal analysis there are many different processing steps occur such as preprocessing, dynamic characteristics, adaptive filtering, parametric estimation, computational intelligence, STFT temporal characteristics, simulation hemodynamics, and cardiac acoustic model estimation.

3.4   CARDIAC SOUND MODELING AND IDENTIFICATION

The central inspiration for model construction is to deepen the knowledge for the physical systems and gain understanding about the physical world. A mathematical model is often realized by a set of equations that review available knowledge and set up rules for how this knowledge interacts. Mathematical models can be developed in different ways: purely theoretically based on the physical relationships which are a priori known about the system, purely empirically by experiments on the already existing system, or by a sensible combination of both ways.

Models obtained by the first method are often called a priori, first principle or theoretical models, while models obtained in the second way are called a posteriori or experimental (black-box) models. In case of theoretical analysis, the dynamic properties of the system are primarily taken care of by the respective balance equations. These balances are established by the laws of conservation supplemented with the necessary state-equations and phenomenological laws. Theoretical model building becomes unavoidable if experiments in the respective plant cannot or must not be carried out. If the plant to be modeled does not yet exist, theoretical modeling is the only possibility to obtain a mathematical model.

Modeling the cardiovascular system requires multivariate, multi-scale, multi-organ integration of information, making it an extremely complex task. The purpose of this section is not to delve into these details but rather to look at two simple models able to reproduce S_1 and S_2. Neither of the models is able to explain the genesis of the heart sounds. However, they do provide adequate representations of the PCG signal, and accordingly, the models can be used to simulate heart sounds.

3.4.1 S_1 CARDIAC SOUND MODELING

The first cardiac sound (S_1) model is somehow precursively approach since the underlying mechanisms of the sound are not fully understood. Based on observations from superficial chest recordings, Chen et al. [64] suggested a model consisting of two valvular components with constant frequency and one myocardial component with instantaneously increasing frequency. The basic idea is that harmonic oscillations associated with atrioventricular valve closure are dampened by the acoustic transmission to the thoracic surface. The valvular components $s_v(t)$ are modeled as a set of transient deterministic signals according to equation 3.1, where N is the number of components, A_i is the amplitude, and ϕ_i is the frequency function of the ith sinusoid.

$$s_v(t) = \sum_{i=1}^{N} A_i(t)sin(\phi_i(t)) \qquad (3.1)$$

The myocardial compartment, coupled with myocardial torsion tension, can be modeled with an modulated amplitude with linear chirp signal according to equation 3.2. $A_m(t)$ is the amplitude modulating wave, $\phi m(t)$ is the frequency function, and s_m is the myocardial component. The frequency of the signal increases during myocardial contraction and levels out as the force plateau is reached [65].

$$s_m(t) = A_m(t)sin(\phi_m(t)). \qquad (3.2)$$

Since the valves close after contraction, the valvular components and the myocardial component are separated by a time delay t_0 before deriving the final S_1 model (equation 3.3). Figure 3.2 shows the spectrogram of an acquired PCG signal of mitral stenosis (MS) or regurgitation case (MR).

$$s_1(t) = s_m(t) + \begin{cases} 0, & \text{if } 0 \leq t \leq t_0; \\ s_v(t - t_0), & t \geq t_0 \end{cases}. \qquad (3.3)$$

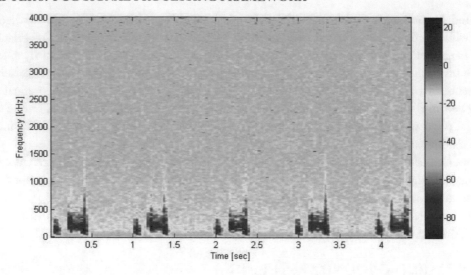

Figure 3.2: Mitral stenosis (MS) spectrogram of an acquired PCG signal.

3.4.2 S_2 CARDIAC SOUND MODELING

Compared to S_1, the underlying mechanisms associated with S_2 are more widely accepted. The aortic component (Ao2) is produced during the closure of the aortic valve while the pulmonary component (Pl2) results from the closure of the pulmonary valve. Each component usually lasts for less than 80 ms. During expiration the two components come closer together (less than 15 ms) while during inspiration, they are separated by 30–80 ms [60]. The separation between the two components is mostly due to different intervals of ventricular systole for the left and the right side of the heart, which is modulated by respiration rhythm as a driving signal for cardiac cycle. As indicated by Bartels et al. [64] and Longhini et al. [65], the resonance frequencies of Ao2 and Pl2 are proportional to the aortic pressure and the pulmonary artery pressure, respectively. This is reasonable since these pressures cause tension in the cardiac structures which affects the frequency of the vibrations. With decreasing pressure in end systole and early diastole, it is thus expected that the instantaneous frequency will decay. According to this hypothesis, A2 and P2 should be composed of short duration frequency modulated transient signals [63], giving an S_2 model consisting of two narrow-band chirp signals; see equation 3.4. A(t) and $\phi(t)$ are instantaneous amplitude and phase functions, and t_0 is the splitting interval between the onset of Ao2 and Pl2:

$$S_2(t) = A_A(t)sin(\phi_A(t)) + A_P(t - t_0)sin(\phi_P(t)) \qquad (3.4)$$

3.4.3   MODELING ABNORMAL HEART SOUND S_3 AND S_4

Physical modeling aims at the quantification of the constitutive properties of cardiac structures (e.g., of the valve leaflets) and the driving forces (e.g., blood pressure). For example, with respect to the second sound, the aortic valve was modeled as a circular elastic membrane, it was allowed to vibrate in interaction with the surrounding blood mass [63, 66].

Typical characteristics of IIA and IIP (respectively, aortic and pulmonary heart sound) could thus be explained, for example, the reduction of amplitude and frequency shift (toward higher frequencies) can be referred as a consequence of the following:

- valve stiffening pathologies;

- diminishing of amplitude in patients with poor ventricular performance (characterized by a slow pressure drop in the ventricle during the isovolumic relaxation);

- augmentation of amplitude in cases of anemia (implying reduced blood viscosity and thus reduced damping in the resonant system).

In a different model, the ventricle is modeled as a finite thick-walled cylinder and the amplitude spectra of computed vibration waveforms contain information concerning the active elastic state of muscular fibers that is dependent on cardiac contractility [27].

Transmission of vibrations by comparing vibrations at the epicardial surface and at the chest wall has been studied [64]. Esophageal PCG (ePCG) proved to be beneficial for recording vibrations originated at the mitral valve [66]. The disappearance of the third sound with aging was explained with the ventricle modeled as a viscoelastic oscillating system with increasing mass during growth [68].

Spectral analysis of the pulmonary component of the second sound reveals information on the pressure in the pulmonary artery [69]. Frequency content and timing of heart vibrations is of major importance; time-frequency analysis of signals is thus performed. Classical Fourier analysis uses harmonic signals (sine and cosine waves) as basic signals. The frequencies of the harmonics are multiples of the fundamental frequency and the signal can be composed by summing the sine and cosine waves multiplied with the Fourier coefficients. Sine waves have an infinite duration and the method is thus beneficial for periodic functions. A phonocardiogram can be considered as a periodic function, but it is composed of a number of phenomena shifted in time with specific frequency content (heart sound components and murmurs). The following MATLAB code displays the initialization and plotting PCG-parameters, in addition to noise function applied to the PCG signals as well. When applying classical Fourier analysis, information on timing is lost. Thus, Fourier analysis has to be performed on shorter time intervals (by dividing the heart cycle into subsequent small intervals) resulting in time and frequency information. Figure 3.3 illustrates the PCG fast fourier transformation (FFT) over the acoustic oscillation spectrum.

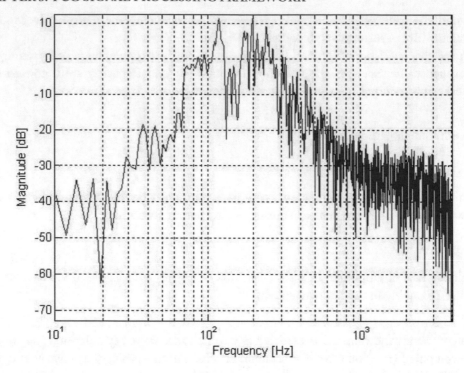

Figure 3.3: Fourier transform of PCG signal illustrates variation of amplitude along a wide spectrum of oscillation frequency.

```
% define PCG signal coefficients and initialization parameters
% p0=0.02; p1=0.52; p2=0.63; p3=1.028;
% w0=p0-0.24; w1=p1-0.172; w2=p2-0.78; w3=p3-0.129;
n=1024;
name='gaussiannoise'; % filtered gaussian noise
name='piece-regular'; % piecewise regular
f=load_signal(name, n);
sigma=0.03*(max(f)-min(f)); % noise level
fn=f+sigma*randn(1,n); % noisy signal
% plot signals
subplot(2,1,1);
plot(f); axis tight; title('Original');
subplot(2,1,2);
plot(fn); axis tight; title('Noisy');
```

To minimize errors resulting from calculating in these small intervals, mathematical techniques have to be applied to overcome the aliasing error originated from lower sampling frequency. Fig. 3.4 shows the various power-spectrum content of different PCG signals in frequency domain.

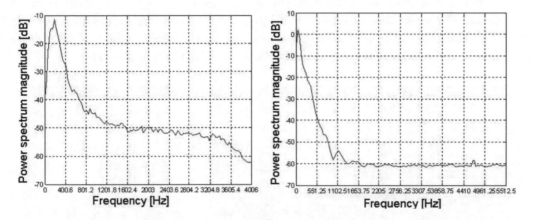

Figure 3.4: Power spectrum of two different PCG signal illustrate different contents of frequency component.

3.5   MODEL-BASED PHONOCARDIOGRAPHY ACOUSTIC SIGNAL PROCESSING

The detection and localization of an acoustic source has long been the motivation of early ultrasonic systems. The need for more precise phonocardiography signal processing techniques has been apparent for quite some time. It has often been contemplated that the incorporation of cardiac acoustic propagation models into signal processing schemes can offer more useful information necessary to improve overall processor performance and to achieve the desired enhancement/detection/localization even under noisy and disturbance conditions. Model-based techniques offer high expectations of performance, since a processor based on the predicted physiological phenomenology that inherently has generated the measured signal must produce a better (minimum error variance) estimate then one that does not. The uncertainty of the myocardial acoustic medium also motivates the use of stochastic models to capture the random and often non-stationary nature of the phenomena ranging from ambient noise and sever murmur conditions. Therefore, processors that do not take these effects into account are susceptible to large estimation errors.

3.6   FUTURE PERSPECTIVES

The extraordinary computational power and miniature size of the microprocessors today may also permit the incorporation of analysis software into the stethoscope for heart sound analysis, or even the provision of diagnosis. Moving toward the goal of automatic diagnosis of heart diseases, computer-based heart sound analysis techniques will continue to evolve. Many advanced signal processing algorithms and data analysis models, for example wavelet transform, adaptive filtering, artificial neural network, and pattern recognition, have already provided new insight into the diagnostic value of heart sound. The exploration of further techniques in the coming years would hopefully help to realize the full potential of cardiac auscultation as a tool for the early detection of heart diseases. Fig. 3.5 shows GUI of labVIEW programming platform for PCG signal acquisition.

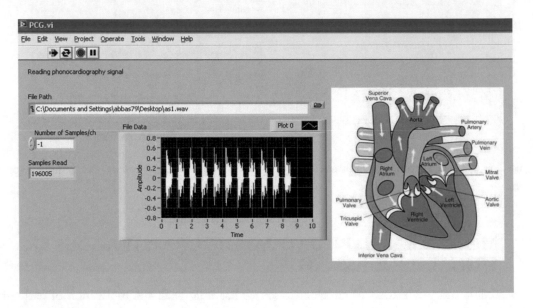

Figure 3.5: Graphical user interface for PCG signal acquisition and analysis based on labVIEW programming platform.

3.6.1   PACEMAKER HEART SOUND ACOUSTIC DETECTION

This approach is considered as a cutting edge technology in the field of intelligent phonocardiography clinical diagnosis module. The basic idea behind this technique was to develop and optimize the mobile auscultation system in order to integrate many of advanced concepts in PCG signal processing and pattern classification with the cardiac pacemaker system, which in proud consist of the last 40 years of PCG signal researches and investigation in this field. The first model prototype with enhanced criteria was introduced in 2005 in PalmMed Trend (Los Angeles), which is a technical symposium for recently invention in medical technology. This system composed of palmtop with windows mobile CE operating system. The PCG signal which involved in integrating cardiac acoustic with pacing therapy control loop, demonstrates robust response for DVI and VVT pacing mode of cardiac pacemaker.

3.6.2   PHYSIOLOGICAL MONITORING OF BLOOD PRESSURE WITH PHONOCARDIOGRAPHY

Blood pressure is an important signal in determining the functional integrity of the cardiovascular system. Scientists and physicians have been interested in blood pressure measurement for a long time. The first blood pressure measurement is attributed to Reverend Stephen Hales, who in the early 18th century connected water-filled glass tubes in the arteries of animals and correlated their blood pressures to the height of the column of fluid in the tubes.

It was not until the early 20th century that the blood pressure measurement was introduced into clinical medicine, albeit with many limitations. Blood pressure measurement techniques are generally put into two broad classes: direct and indirect. Direct techniques of blood pressure measurement were invasive techniques, involving a catheter being inserted into the vascular system. The indirect techniques are noninvasive, with improved patient comfort and safety, but at the expense of accuracy. The accuracy gap between the invasive and the noninvasive methods, however, has been narrowing with the increasing computational power available in portable units, which can crunch elaborate signal processing algorithms in a fraction of a second. During a cardiac cycle, blood pressure goes through changes, which correspond to the contraction and relaxation of the cardiac muscle, with terminology that identifies different aspects of the cycle.

The maximum and minimum pressures over a cardiac cycle are called the systolic and diastolic pressures, respectively. The time average of the cardiac pressure over a cycle is called the mean pressure, and the difference between the systolic and diastolic pressures is called the pulse pressure. Normal blood pressure varies with age, state of health, and other individual conditions. An infant's typical blood pressure is 80/50 mmHg (10.66/6.66kPa) (systolic/diastolic). The normal blood pressure increases gradually and reaches 120/80 (15.99/10.66 kPa) for a young adult. Blood pressure is lower during sleep and during pregnancy. Many people experience higher blood pressures in the medical clinic, a phenomenon called the white coat effect.

In the traditional, manual, indirect measurement system, an occluding cuff is inflated and a stethoscope is used to listen to the sounds made by the blood flow in the arteries (Fig. 3.6), called Korotokov sounds. When the cuff pressure is above the systolic pressure, blood cannot flow, and no sound is heard. When the cuff pressure is below the diastolic pressure, again, no sound is heard.

A manometer connected to the cuff is used to identify the pressures where the transition from silence to sound to silence is made. This combination of a cuff, an inflating bulb with a release valve, and a manometer is called a sphygmomanometer and the method an auscultatory technique. Usually, the cuff is placed right above the elbow, elevated to the approximate height of the heart, and the stethoscope is placed over the brachial artery. It is possible to palpate the presence of pulse under the cuff, rather than to use a stethoscope to listen to the sounds. The latter approach works especially well in noisy places where it is hard to hear the heart sounds.

The use of the PCG signal to identified blood pressure information was a newly developed method which is used recently as an alternative to direct oscillometric blood pressure method. The first adaptable technique was developed by Cornell et al. [2004]. The wavelet decomposition system

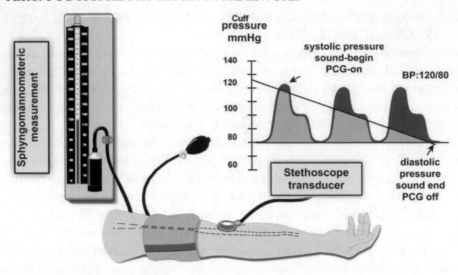

Figure 3.6: Oscillometric blood pressure measurement procedure which integrate with the phonocardiography to optimize the non-invasive blood pressure measurement.

was used for separation and identification of the blood pressure and hemodynamic information from the phonocardiography traces.

This method has various sources of potential error. Most of these sources are due to misplacement of the cuff, problems with hearing soft sounds, and the using of wrong cuff size. Using a small cuff on a large size arm would result in overestimating the blood pressure, and vice versa. Nevertheless, an auscultatory measurement performed by an expert health care professional using a clinical graded sphygmomanometer is considered to be the gold standard in noninvasive measurements.

3.6.3   AUTOMATED BLOOD PRESSURE-PCG BASED MEASUREMENT

In the earlier measurement units, it was a combination of hardware and software that controlled the various aspects of the automated measurement (or estimation) of blood pressure. With the increasing computational power of micro-controllers units (MCU), all decision making and control are now implemented in software and with more elaborate algorithms. Here are some functions that are included in a measurement system. Refer to Fig. 3.7 for a typical organization of such algorithms and application in blood pressure measurement system.

- **Inflation/deflation control**: Whether data collection is done during inflation or deflation, continuously or in steps, there are many challenges to appropriately control the air pump. This include maintaining a smooth baseline cuff pressure without filtering out the arterial variations; adjusting the pump speed to variations arising from different cuff sizes, arm sizes, and cuff tightness; and selecting the range of cuff pressures for which data will be collected.

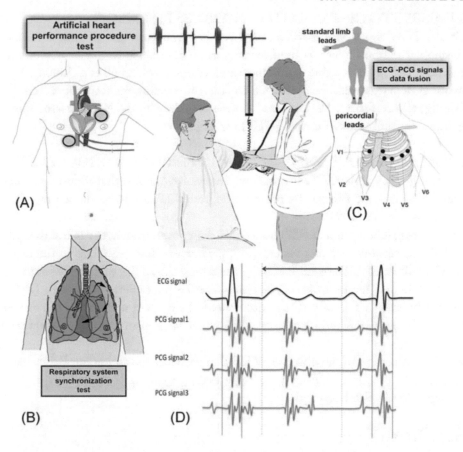

Figure 3.7: Phonocardiography signal processing used in many cardiovascular clinical application, where (A) is application of heart sound in performance test of left-ventricular assistive device (LVAD), (B) Use of PCG in the respiratory system synchronization test, (C) Application of PCG signal with ECG-acquisition for a physiological system modeling and correlation. and (D) is the ?????.

- **Pulse detection**: This is a fundamental part of extracting features from raw cuff pressure data. It becomes especially challenging when conditions, such as arrhythmia, or tremors, affect the regularity of pulses. Pattern recognition techniques with features found in time, frequency, or wavelet domains are used to deal with difficult situations.

- **Blood pressure estimation**: The indirect method of measurement is a process of estimating pressures with the use of features extracted from cuff-pressures or other transducer data. This algorithm used to be limited to linear interpolation. Recently, more elaborate decision-making and modeling tools such as nonlinear regression, neural networks, and fuzzy logic are also being used for this purpose.

3.6.4 TRANSIT TIMES EXTRACTION AND ESTIMATION

The use of the PCG signal to identified blood pressure and pulse transit time information was a newly developed method which is used recently as an alternative to direct oscillometric blood pressure method. The first adaptable technique was developed by Cornell et al. [2004]. The wavelet decomposition system was used for extraction the spatial information and the temporal characteristics of the PCG-signal waveform to be correlated in time domain with the blood pressure profile to set up timing-frame for pulse transient time (PTT) paradigm.

3.6.5 HEMODYNAMICS AND TRANSIT INTERVALS MODULATION

Due to the technique by which PCG is measured, studies have shown that time-related parameters derived from PCG are not affected by pre-ejection period (PEP) as conducted by Hasegawa et al. (1991) and Visser et al. (1993).

As monitoring blood pressure (BP) response during postural changes can test the sympathetic functions of the autonomic nervous system [20], parameters that can relate to the observed BP changes are clinically useful. The examination of the typical time delay value between the first heart sound (S_1) of PCG and the upstroke of the corresponding photoplethysmography (PPG) signals or vascular transit time (VTT) of healthy adults should be done in addition to the determination of the effect of hemodynamic turbulences on VTT when the measured periphery adopts different postures.

Also, the association of the observed VTT changes with corresponding BP changes should be considered in the computation. Finally, the regression of equations to relate VTT and BP during postural changes should also be considered.

3.7 SUMMARY

The phonocardiography signal processing (Fig. 3.7) plays a vital role in development intelligent stethoscopic system as an integrated approach for smart auscultation technique.

The signal processing methods vary in its final application field where for spatially based signal processing the resulted processed PCG signal can be used in automated diagnosis of cardiac valve insufficiency, and this method can be further integrated to be a clinical diagnostic tool.

Key points can be summarized as follows:

- PCG signal processing fundamentals;

- Fourier transform of PCG signal;

- Temporal analysis technique, frequency-adapted signal processing, and representation;

- Spectral analysis of PCG signal;

- Electronic cardiac stethoscope signal processing aspects, including pre-filtering, signal buffering, Fourier transform, and signal presentation.

CHAPTER 4

Phonocardiography Wavelets Analysis

4.1 WAVELETS

Wavelets have been found to be very useful in many scientific and engineering applications, including signal processing, communication, video and image compression, medical imaging, and scientific visualization. The concept of wavelets can be viewed as a synthesis of ideas that originated during the last several decades in engineering, physics, and pure mathematics. However, wavelet is a rather simple mathematical tool with a great variety of possible applications. The subject of wavelets is often introduced at a high level of mathematical sophistication. The goal of this chapter is to develop a basic understanding of wavelets, their origin, and their relation to scaling functions, using the theory of multi-resolution analysis.

4.1.1 HISTORICAL PERSPECTIVE

Prior to the 1930s, the main tools of mathematics for solving scientific and engineering problems traced back to Joseph Fourier (1807) with his theory of frequency analysis. He proposed that any 20-periodic function f(t) can be represented by a linear combination of cosines and sines components:

$$f(t) = a_0 + \sum_{k=1}^{\inf} (a_k cos(kt) + b_k sin(kt)). \tag{4.1}$$

The coefficients a_0, a_k, b_k are the Fourier coefficients of the series and are given by

$$a_0 = \frac{1}{2\pi} \int f(t)dt. \tag{4.2}$$

$$a_k = \frac{1}{2\pi} \int f(t)cos(kt)dt. \tag{4.3}$$

$$b_k = \frac{1}{2\pi} \int f(t)sin(kt)dt. \tag{4.4}$$

After 1807, mathematicians gradually were led from the notion of frequency analysis to the notion of scale analysis that is, analyzing f(t) by creating a mathematical structure that varies in scale. A. Haar, in his thesis (1909), was the first to mention the using of wavelets. An important property of the wavelets he used is that they have compact support, which means that the function vanishes outside a finite interval. Unfortunately, Haar wavelets are not continuously differentiable, which limits their application. From 1930s–1960s, several groups, working independently, researched the representation of functions using scale-varying basis functions. By using one such function, the Haar basis function, Paul Levy investigated Brownian motion and thereby laid the foundation for the modern theory of random processes. Levy found that the Haar basis function is superior to the Fourier basis functions for studying small and complicated details in Brownian motion. Also during the 1930s, research was done by Littlewood et al. [73] on computing the energy of a time-function f(t):

$$energy = \frac{1}{2\pi} \int |f(t)|^2 \, dt. \tag{4.5}$$

Their computation produced different results when the energy was concentrated around a few points and when it was distributed over a larger interval. This observation disturbed many scientists, since it indicates that energy might not be conserved. Later on, they discovered a function that can both vary in scale and conserve energy at the same time, when computing the functional energy. David Maar applied this work in developing an efficient algorithm for numerical image processing using wavelets in the early 1980s. Between 1960 and 1980, the mathematicians Guido Weiss and Ronald Coifman studied the simplest elements of a function space, called atoms, with the goals of finding the atoms for a common function and finding the construction rules that allow the reconstruction of all the elements of the function space using these atoms. In 1980, Grossman and Morlet [71] recasted the study of quantum physics in the context of wavelets using the concept of frames. Morlet introduced the term wavelets as an abbreviation of wavelet of constant shape. These new insights into using wavelets provided a totally new way of thinking about physical reality. In 1985, Stephane Mallat applied wavelets to his work in digital signal processing. He discovered a relationship between quadrature mirror filters, the pyramid algorithm, and orthonormal wavelet bases. Inspired by these results, Y. Meyer constructed the first nontrivial wavelets. Unlike the Haar wavelets, the Meyer wavelets are continuously differentiable; however, they do not have compact support. In the early 1990s, Ingrid Daubechies used Mallat's work to construct a set of orthonormal wavelet basis functions that are perhaps the most elegant, and have become the cornerstone of wavelet applications today.

The development of wavelets is an emerging field comprising ideas from many different fields. The foundations of wavelet theory have been completed, and current research is in the refinement stage. The refinement involves generalizations and extensions of wavelets, such as extending wavelet packet techniques. The future of wavelets depends on the possibility of applications. Wavelets decomposition (Fig. 4.1) have so far been limited in practical applications by their lack of compact support.

4.2   FOURIER ANALYSIS

Time-series data have traditionally been analyzed in either the time or the frequency domain. Fourier analysis is quite useful in identifying frequency components of a signal, but it cannot describe when those frequency components occurred, since it lacks time resolution. This is particularly important for signals with time-varying frequency content, as in human speech and video images.

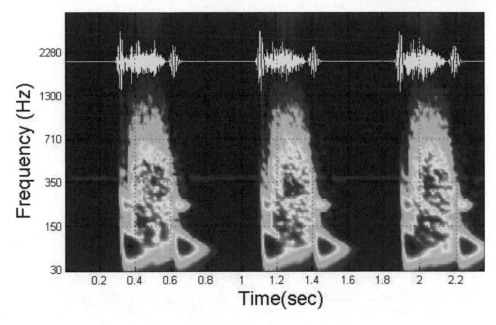

Figure 4.1: Phonocardiography wavelets transformation based on biorthogonal signal decomposition.

The Fourier transform is characterized by the ability to analyze a signal in the time domain for its frequency content. The transform works by first translating a function in the time-domain into a function in the frequency domain. The signal can then be analyzed for its frequency content, because the Fourier coefficients of the transformed function represent the contribution of each sine and cosine function at each frequency.

An Inverse transform does the opposite by transforming data from the frequency domain into the time domain. Although the time-series data can have infinitely many sample points, in practice one deals with a finite time interval using a sampling mechanism. The Discrete Fourier Transform (DFT) estimates the Fourier transform of a function from a finite number of its sampled points. The sampled points are supposed to be typical of what the signal looks like at all other times.

Figure 4.2 shows the Fourier analysis of phonocardiography, the phase trace of it and the FFT-amplitude of the same PCG signal. The DFT has symmetry properties almost exactly the same as the continuous Fourier transform. Approximation of a function by samples, and approximation of the Fourier integral by the DFT, requires multiplication by a matrix which involves on the order of (n) arithmetic operations. However, if two samples are uniformly spaced, then the Fourier matrix can be factored into a product of just a few sparse matrices, and the resulting factors can be applied to a vector in a total order of n log(n) arithmetic operations. This technique is the so-called Fast Fourier Transform (FFT).

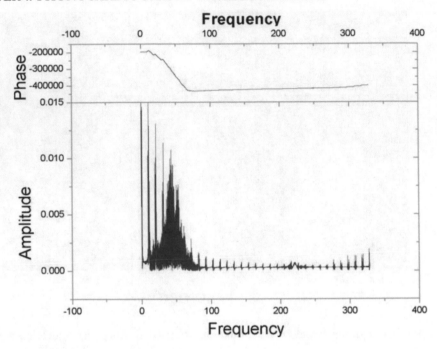

Figure 4.2: Fourier analysis of phonocardiography (above) the phase of PCG trace (below). The FFT-amplitude of the same PCG signal.

4.2.1   WAVELETS VERSUS FOURIER ANALYSIS

The FFT and the discrete wavelet transform (DWT) are both linear operations that generate a data structure containing (n) segments of various lengths, usually filling and transforming them into a different data vector of length 2^n. The mathematical properties of the matrices involved in the transforms are similar as well.

The inverse transform matrix for both the FFT and the DWT is the transpose of the original. As a result, both transforms can be viewed as a rotation in function space to a different domain [68]. For the FFT, the new domain contains basis functions that are sines and cosines. For the wavelet transform, the new domain contains more complicated basis functions called wavelets, mother wavelets, or analyzing wavelets. Both transforms have another similarity; the basis functions are localized in frequency, making mathematical tools such as power spectra (how much power is contained in a frequency interval) and scalograms useful at picking out frequencies and calculating power distributions.

The most interesting dissimilarity between these two kinds of transforms is that individual wavelet functions are localized in space, while Fourier sine and cosine functions are not. This localization in space, along with wavelets localization in frequency, makes many functions and operators using wavelets sparse when transformed into the wavelet domain. This sparseness, in turn, makes wavelets useful for a number of applications such as data compression, feature detection in images, and noise removal from time series. One way to see the time frequency resolution difference between the two transforms is to look at the basis function coverage of the time frequency plane [64, 65].

The square-wave window truncates the sine or cosine function to particular width. Because a single window is used for all frequencies in the WFT, the resolution of the analysis is the same at all locations in the time frequency plane. An advantage of wavelet transform is that the windows vary. In order to isolate signal discontinuities, one would like to have some very short basis functions. At the same time, in order to obtain detailed frequency analysis, one would like to have some very long basis functions.

A way to achieve this is to have short high-frequency basis functions and long low-frequency ones. This medium is exactly what you get with wavelet transform and this is shown in Fig. 4.3, in which it gives the illustration of continuous wavelet analysis (CWT) based on Haar transformation function for a three cases (normal, valve regurgitation and stenosis).

4.2.2   HAAR WAVELET

The Haar scaling function and Haar wavelet are a very simple example to illustrate many nice properties of scaling functions and wavelets, and are of practical use as well. The Haar scaling function is defined by:

$$\phi(t) = \begin{cases} 1 & 0 \le x \le 1 \\ 0, & \text{otherwise} \end{cases}. \tag{4.6}$$

The two-scale relation can be expressed as a summation:

$$\Phi_2^H = \sum_{k=0}^{2} p_k \phi(2t - k) \tag{4.7}$$

or

$$\phi(t) = \phi(2t) + \phi(2t - 1) \tag{4.8}$$

The Haar wavelet corresponding to the Haar scaling function is given by:

$$\Psi(t) = \begin{cases} -1 & for & 0 \le x \le \frac{1}{2} \\ -1 & for & \frac{1}{2} \le x \le 1 \\ 0 & for & \text{otherwise} \end{cases}. \tag{4.9}$$

The construction of the two-scale relation for the Haar wavelet is easily computed as follows:

$$\psi(2t) = \phi(2t) - \phi(2t - 1). \tag{4.10}$$

The two-scale relations express $\phi(t)$ in terms of $\phi(2t)$ and $\phi(2t - 1)$, while the other two-scale relations are for Haar wavelets express $\psi(t)$ in terms of $\phi(2t)$ and $\phi(2t - 1)$. The reconstruction relations can be written in the matrix form:

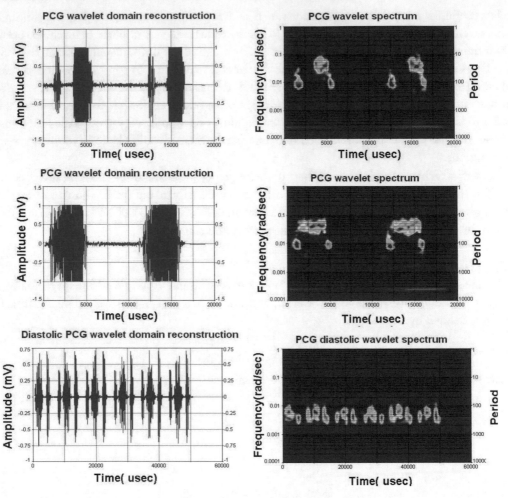

Figure 4.3: PCG signal wavelets analysis of different cardiac valvular disorder (right) normal cardiac murmurs, (middle) mitral regurgitation, (left) mitral stenotic PCG trace, and the corresponding wavelet derived spectrum for each cases.

$$\begin{bmatrix} \phi(t) \\ \psi(t) \end{bmatrix} = \begin{bmatrix} 1 & 1 \\ -1 & 1 \end{bmatrix} = \begin{bmatrix} \phi(2t) \\ \phi(2t-1) \end{bmatrix}. \tag{4.11}$$

The decomposition relations are easily derived by just inverting the reconstruction relations as follows:

$$\begin{bmatrix} \phi(2t) \\ \pi(2t-1) \end{bmatrix} = \begin{bmatrix} \frac{1}{2} & \frac{1}{2} \\ -\frac{1}{2} & \frac{1}{2} \end{bmatrix} = \begin{bmatrix} \phi(t) \\ \psi(t) \end{bmatrix}. \tag{4.12}$$

4.2.3 DEBAUCHIES (DB) WAVELET

Another example of wavelets defined on the real line is Daubechies wavelets in which Figs.4.4 and 4.5 show Db-wavelet transformation of a different PCG signals. The Daubechies wavelets show a robust result in the analysis of linear time-varying biomedical signals, due to it ability to compensate for missed coefficient in the final decomposition paradigm. The ϕ_3^D is the Daubechies scaling function, and it is defined by the following relation:

$$\phi_3^D = \sum_{k=0}^{3} p_k \phi(2t - k) \tag{4.13}$$

where two-scale sequence p_k are expressed as below:

$$p_1, p_2, p_3, p_4 = \left\{ \frac{1 + \sqrt{3}}{4}, \frac{3 + \sqrt{3}}{4}, \frac{3 - \sqrt{3}}{4}, \frac{1 - \sqrt{3}}{4} \right\}. \tag{4.14}$$

The normalized process of heart sound signal, and (N) was signal length of 20 ms. An initial threshold was set to 0.2 times of the maximum amplitude to identify S_1 and S_2. Then, a limit window of 50 ms was slid throughout the whole set of the average Shannon energy signals. The threshold was adjusted to attain the required peaks. Based on the R-R intervals and by observing the pause lengths, S_1 and S_2 were then identified separately.

Thus, the acquired signals were segmented into cycles. It is known that pathological murmurs are commonly heard in patients with cardiac abnormalities such as valvular disease mitral regurgitation (MR cases), shunts or narrowed vessels [67]. Based on this understanding, the additional detectable heart sound components in each segmented cycle using the segmentation procedures were considered as murmurs.

Performance of heart sound signal characterization depends on PCG component identification and its characteristics, which is based on PCG signal classification and the threshold of heart cycle. Shannon energy was adjusted until all possible components were identified. Thereafter, the amplitude, frequency, time-span, and interval (including systolic and diastolic intervals) should determined. In general, the two-scale sequence for any PCG scaling functions has the property as presented below:

$$\sum_{k} p_{2k} = \sum_{k} p_{2k-1} = 1. \tag{4.15}$$

There is no closed form for ϕ_3^D. The Haar wavelet is the simplest one; it has many ϕ_3^D applications. However, it has the drawback of discontinuity. It consists entirely of rectangular functions and cannot reproduce even linear functions smoothly in finite series for practical use. On the other hand, B-spline wavelets have higher continuity than Haar wavelets. They are more suitable for representing any continuous function. However, the complications of calculating its wavelet decomposition and reconstruction relation coefficients have limited its usefulness. The MATLAB code for wavelet analysis is shown below.

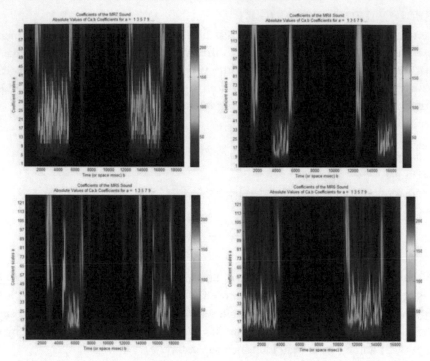

Figure 4.4: Application of Db-Wavelet transformation on PCG signal for different clinical cases.

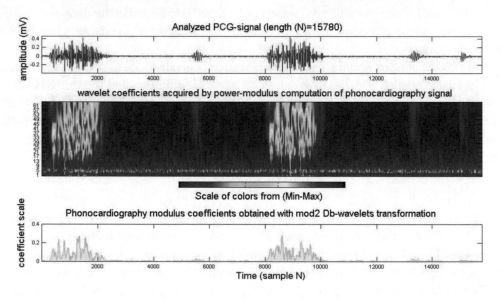

Figure 4.5: Db-wavelet AS (aortic stenosis) phonocardiography signal decomposition with six-level denoising.

```
clf
t=0:.01:5;
% define the PCG signal input vector
xt=cos(10*t)+2; %This is the unsampled PCG input to the IIR digital filter module
n=0:50;
Ts=.1;
xn1=2;%This signal should pass through the filter
xn2=cos(10*n*Ts);%This signal should not pass
xn=xn1+xn2;
subplot(3,1,1); plot(t, xt);
title('The input signal:cos(10*t)+2 before filtering');
for N=2:4:6
%digital filter with orders 2 and 6
[numz denz]=butter(N, 0.1592)%The transfer function of the digital filter
[H, f]=freqz(numz, denz);
subplot(3,1,2); plot(f/pi, abs(H)); hold on;
% apply continuous wavelet decomposition
% load the PCG signal under test
PCG=waveread('pcg/_data/mitral regurgitation/sample/mr1.wav');
[p,s]=wavedec(PCG,lev,wname);
scales = (1:32); % Levels 1 to 5 correspond to scales 2, 4, 8, 16 and 32.
cwt(PCG,scales,wname,'plot');
colormap(pink(nbcol))
lev = 5;
wname = 'db1';
nbcol = 64;
len = length(PCG);
cfd = zeros(lev,len);
for k = 1:lev
d = detcoef(p,s,k);
d = d(:)';
d = d(ones(1,2^k),:);
cfd(k,:) = wkeep1(d(:)',len);
end
cfd = cfd(:);
I = find(abs(cfd)<sqrt(eps));
cfd(I) = zeros(size(I));
cfd = reshape(cfd,lev,len);
cfd = wcodemat(cfd,nbcol,'row');
```

```
%
%recomputing the continuous wavelet transformation
%for the frequency mapped signal;
%
set(subplot(3,1,1),'Xtick',[]);
plot(PCG,'r');
title('Analyzed signal.');
set(gca,'Xlim',[1 length(PCG)])
subplot(3,1,2);
colormap(cool(128));
image(cfd);
tics = 1:lev;
labs = int2str(tics');
set(gca,'YTicklabelMode','manual','Ydir','normal',
'Box','On','Ytick',tics,'YTickLabel',labs);
title('Discrete Transform, absolute coefficients.');
ylabel('Level');
% CWT at debauchies transform with 5th-level
set(subplot(3,1,2),'Xtick',[]);
subplot(3,1,3);
scales = (1:64);
cwt(PCG,scales,'db5','plot');
colormap(cool(128));
tt = get(gca,'Yticklabel');
[r,c] = size(tt);
yl = char(32*ones(r,c));
for k=1:3:r;
yl(k,:) = tt(k,:);
end
set(gca,'Yticklabel',yl);
```

4.2.4   SUBBAND CODING

One of the main applications of subband coding is compression. A key concept in signal analysis is that of localization in time and frequency. Another important intuitive concept is that of multi-resolution, or the idea that one can consider a signal at different levels of resolution. These notions are particularly evident in image processing and computer vision, where coarse versions of images are often used as a first approximation in computational algorithms. In signal processing, a low-pass and sub-sampled version is often a good coarse approximation for many real-life signals. This intuitive paradigm leads to the mathematical framework for wavelet constructions [72]. The wavelet decomposition is a successive approximation method that adds more and more projections onto detail spaces, or spaces spanned by wavelets and their shifts at different scales.

In addition, this multi-resolution approximation is well suited to many applications. That is true in cases where successive approximation is useful; for example, in browsing through image databases, as is done for instance on the World-Wide Web. Rather than downloading each full image, which would be time consuming, one only needs to download a coarse version, which can be done relatively fast. Then, one can fetch the rest, if the image seems of interest.

Similarly, for communication applications, multi-resolution approximation leads to transmission methods, where a coarse version of a signal is better protected against transmission errors than the detailed information. The assumption is that the coarse version is probably more useful than the detail.

There are many techniques for image coding; the subband coding method is the most successful today because it gives a fine details of the signal over a wide range of frequencies. The pyramid coding is effective for high-bit-rate compression, while transform coding based on the discrete cosine transform has become the Joint Photographer Expert Group (JPEG) standard. The subband coding using wavelets transform (the tree-structured filter-bank approach) avoids blocking at medium bit rates, because its basis functions have variable length. It uses an adapted basis (the transformation depends on the signal). Long basis functions represent fat background (low frequency), and short basis functions represent regions with texture.

This feature is good for image enhancement, image edge detection, image classification, video-conferencing, video on demand, tissue, and cancer cell detection [74]. Due to its adapted basis functions, one can also develop a set of algorithms for adaptive filtering systems [75].

4.3   WAVELETS DECOMPOSITION

A wavelet allows one to do multi-resolution analysis, which helps to achieve both time and frequency localization. Here, the scale (or resolution, actually it is the inverse of frequency) that we use to look at data plays a vital role. Wavelet algorithms process data at different scales or resolutions.

Figure 4.6 shows the decomposition of PCG signal based on wavelets approach. If we look at a signal with a large (window), we would notice gross (or averaged) features. Similarly, if we look at a signal with a small (window), we would notice detailed features. Thus, by using varying resolution, the problem that was there with Short Time Fourier Transform (STFT) will be solved, due to the use of fixed window size (or resolution).

The core of a wavelet analysis procedure is the choice of a wavelet prototype function, called a mother wavelet. Temporal analysis is performed with a contracted, high-frequency version of the prototype wavelet, while frequency analysis is performed with a dilated, low-frequency version of the same wavelet.

Because the original signal can be represented in terms of a wavelet expansion (using coefficients in a linear combination of the wavelet transform), data operation can be performed using just the corresponding wavelet coefficients.

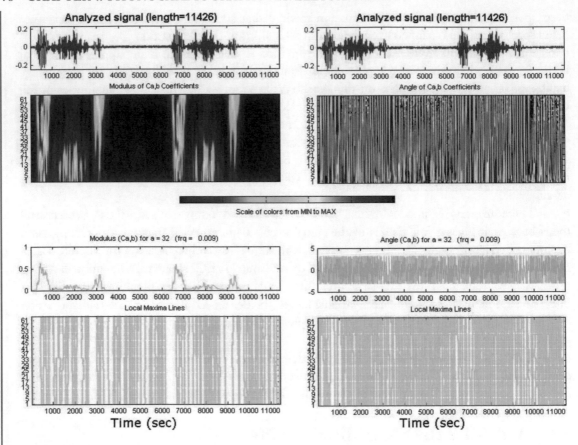

Figure 4.6: Decomposition of PCG signal based on wavelets approach.

4.3.1 CONTINUOUS WAVELET TRANSFORM

Continuous wavelet transform can be formally written as:

$$\gamma(s, \tau) = \int f(t) \Psi_{s,\tau}^*(t) dt. \tag{4.16}$$

The * denotes complex conjugation. This equation shows how a function f(t) is decomposed into a set of basis functions s; $\Psi(t)$, called wavelets. The variables s and τ, scale and translation, are the new dimensions after the wavelet transform. The wavelets are generated from a single basic wavelet (t) (that satisfies some conditions like admissibility etc.), the so-called mother wavelet, by scaling and translation

$$\Psi_{s,\tau} = \frac{1}{\sqrt{s_1}} (\frac{t-s}{s}). \tag{4.17}$$

Here, (s) is the scale factor, s_1 is the translation factor and the factor $s_1 = 2$ is used for energy normalization across the different scales. It should be noted that in the above equations the wavelet basis functions are not specified. This is the main difference between the Fourier transform and the wavelet transform. The theory of wavelet transforms deals with the general properties of wavelet. Thus, it defines a framework, based on which one can design the wavelet he wants. Figure .4.7 shows different wavelet transformation procedures such as Morlet, Meyer, Gaussian, and Mexican hat.

4.3.2   DISCRETE WAVELET TRANSFORM

The continuous wavelet transform (CWT) described in the last section has redundancy. CWT is calculated by continuously shifting a continuously scalable function over a signal and determining the correlation between them. It is clear that these scaled functions will be near an orthonormal basis and the obtained wavelet coefficients will therefore be highly redundant. To remove this redundancy discrete wavelet transform (DWT) (Fig. 4.8) is used.

In DWT the scale and translation parameters are chosen such that the resulting wavelet set forms an orthogonal set, i.e., the inner product of the individual wavelets are equal to zero. Discrete wavelets are not continuously scalable and translatable but can only be scaled and translated in discrete steps. This is achieved by modifying the wavelet representation as:

$$\Psi_{s,\tau} = \frac{1}{\sqrt{s_0^s}} \Psi \left(\frac{t - \tau \tau_0 s_0^s}{s_0^s} \right). \tag{4.18}$$

Here, s and τ are integers and s_0^s is a fixed dilation step. τ_0 is the translation factor and it depends on the dilation step. The effect of discretizing the wavelet is that the time-scale space is now sampled at discrete intervals. When the $s_0 = 2$, the sampling of the frequency axis corresponds to dyadic sampling will choose. For the translation factor we generally choose $\tau_0 = 1$. In that case, the previous equation becomes:

$$\Psi_{s,\tau} = \frac{1}{\sqrt{2^s}} \Psi \left(\frac{t - \tau 2^s}{2^s} \right). \tag{4.19}$$

Table 4.1 presents the identified percentage of S_1 and S_1 waveforms in the PCG data set and the corresponding energy value and entropy Φ_{PCG} based on adaptive wavelet analysis. As indicated in this table, the minimum error value is 0.0021 and the maximum one is 0.0051 which show an acceptable range of biased error during classification process of PCG signals.

In comparison with the other wavelet decomposition method the adaptive Db-wavelet shows a considerable degree of superiority among other wavelet decomposition technique such as Haar and biorthogonal methods.

4.4   PATTERN DETECTION BASED ON ADAPTIVE WAVELETS ANALYSIS

The principle for designing a new wavelet for CWT, is to approximate a given pattern using Least squares Optimization (LEO), as the results of this wavelets analysis illustrated in Fig. 4.9, where

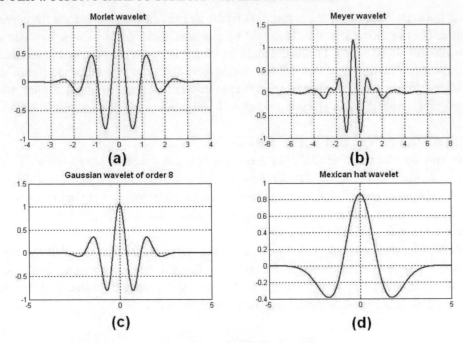

Figure 4.7: Wavelet transformation family (a) Morlet, (b) Meyer, (c) Gaussian, (d) Mexican hat.

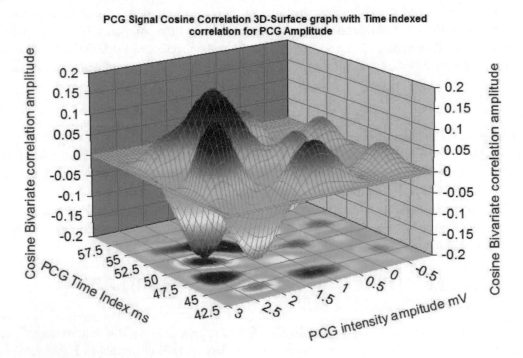

Figure 4.8: 3D phonocardiography signal decomposition based on DWT-method.

Table 4.1: PCG signal analysis based on wavelets decomposition with the entropy for PCG intensity profile.

Energy (mW)	Entropy (Φ_{PCG})	Biased error	S_1-% predicted	S_2-% predicted	Identified PCG%	p-value	SIR index	Mean value
203.66	1.293	0.0042	92.31%	92.61%	83.26± 0.13	0.0031	1.682	14.92
197.87	1.312	0.0023	91.62%	92.03%	95.51± 0.34	0.0028	1.732	16.38
193.42	1.352	0.0045	92.07%	92.12%	91.74± 0.32	0.0022	1.788	14.27
196.21	1.392	0.0051	90.36%	90.22%	93.90± 0.46	0.0034	1.892	15.29
191.93	1.421	0.0039	89.81%	90.94%	92.04± 0.32	0.0028	2.013	16.33
191.25	1.469	0.0036	91.02%	90.73%	93.09± 0.41	0.0032	1.703	16.51
190.01	1.532	0.0021	90.08%	90.01%	87.48± 0.29	0.0034	1.841	15.07
191.05	1.362	0.0028	89.05%	90.85%	88.87± 0.96	0.0022	2.451	14.86

under some constraints leading to an admissible wavelet; which is well suited for the pattern detection using the continuous wavelet transform.

The adaptive pattern detection have a prospect application in phonocardiography pattern localization with wavelets decomposition.

As Fig. 4.5 shows previously, the localization of the cardiac hemodynamic events during aortic stenosis (AS) based on adaptive wavelet algorithm, is more obviously and delineated results other than wavelet decomposition algorithm. In spite, the Db-wavelets algorithm also shows stability and robustness in identification and extracting of hemodynamic information from the PCG signal.

Referring to other research work in adaptive pattern detection of cardiac event as automated annotation, which could help for fusion of the two clinical cardiac valve investigation methods (e.g., echocardiography and adaptive PCG pattern detection) to perform a linear and efficient diagnostic interpretation and analysis. The use of Shannon wavelet template, as shown in Fig. 4.10, in PCG decomposition will improve the results obtained to some extents [76].

The wavelet decomposition of PCG vector can be done as in the following equation:

$$S(t, w) = \int z(t + \tau/2)z^*(t - \tau/2)e^{i\omega t}d\tau \tag{4.20}$$

where

$$z(t) = s(t) + i.H(t) \tag{4.21}$$

where H(t) is the imaginary part of wavelet transformation function.

The output PCG decomposed profile updated as follows:

$$S(t, w, \phi) = \int_{k=0}^{\phi=K} z(t + \tau/2)z^*(t - \tau/2)e^{i\omega\phi t}d\tau + d\phi. \tag{4.22}$$

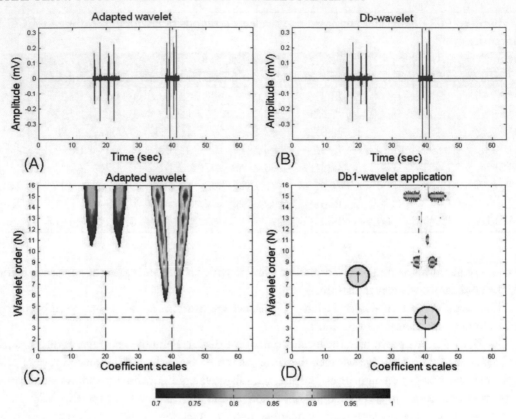

Figure 4.9: Phonocardiography adaptive wavelet pattern detection based on Least square optimization (LEO) algorithm, where (A) temporal representation of PCG- adaptive wavelet function; (B) is Db-wavelet function; (C) scaled coefficients of adaptive wavelet function; (D) the scale coefficients plot of Db-wavelet function.

The systematic continuous transformation can also used for synthesis the new wavelets template to be embedded in the adaptive pattern localization:

$$S(t, w) = \int \int \frac{1}{\sqrt{\tau^2/\sigma}} s(u + \tau/2) s^*(u - \tau/2) e^{[(u-t)^2/(4\tau^2/sigma)]-i\omega t} d\tau du. \qquad (4.23)$$

However, some of the results obtained through biorthogaonal decomposition (Bior-wavelet methods) show acceptable results to some degree, the Debauches (DbW) and adaptive wavelets (AWT) are more robust in dynamic PCG signal analysis.

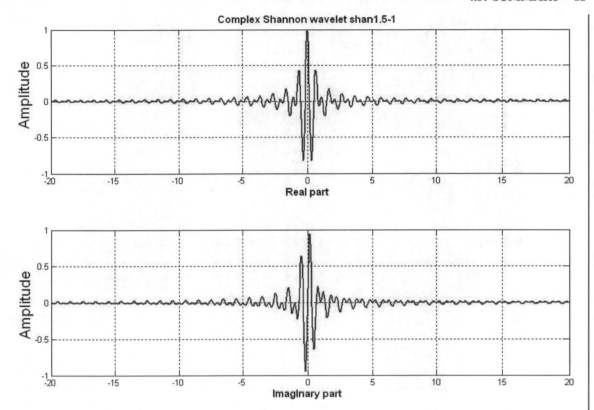

Figure 4.10: Shannon wavelet decomposition template used for optimization pattern detection of PCG signal.

4.5   SUMMARY

In the previous sections, the wavelet analysis of phonocardiography signal was discussed. Different approaches of wavelet transformations including (Haar, Debauches, Bi-orthogonal, and STFT) were touched in this chapter, in addition to the use of continuous wavelet transformation in the analysis, classification, and identification of a varieties of heart sounds traces.

The wavelets is considerably a powerful tool in biomedical signal analysis and processing. The adaptive wavelets analysis and decomposition were also illustrated, in which the combined PCG-signal wavelet decomposition and reconstruction is of stable methods for localization and investigation of the temporal characteristics and related spectral and spatial parameters.

As a summary of this chapter, the following assumptions and remarks can be helpful for further advances in wavelets adaptive signal processing and other wavelet derivatives algorithm.

- The use of higher-order wavelets to perform more complex pattern classification and spectral analysis of PCG signals.

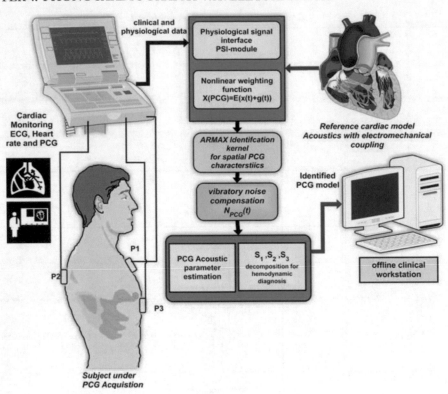

Figure 4.11: Block diagram of the PCG ARMAX system identification, the computational loop consist of clinical monitoring system (Ohmeda Medical. Inc. USA) which is connected to the analysis workstation of Windows-based platform. The main kernel consist of the (1) nonlinear weighting function, (2) ARMAX-identification

- Developing a hybrid wavelet algorithm with other computational intelligence and supervisory data clustering techniques for synthesis a robust and high-performance PCG data mapping and recognition.

CHAPTER 5

Phonocardiography Spectral Analysis

5.1 PCG SIGNAL SPECTRAL ANALYSIS

Heart sounds are complex and highly non-stationary signals in their nature and have been known to be quasi-stationary signals for a long time. The "heart beats" associated with these sounds are reacted in the signal by periods of relatively high activity and rhythmic energy style, alternating with comparatively intervals of low activity. Accordingly, PCG Spectrometric properties can be extracted by different methods using (e.g., Short-Time Fourier Transformation (STFT)), as it estimates the power spectral density (PSD) of successive waveform and computed these transformation will lead to periodic estimation of energy spikes within the acoustical waveform.

The PCG spectral analysis will be considered in this chapter; treat of, determining the distribution in frequency of the power of a time series signal from a finite set of measurements. Spectral analysis has found wide applications in diverse fields such as, radar, sonar, speech, biomedicine, economics, geophysics, and others; in which the spectral contents of signals are of interest. For example, in radar and sonar systems, the locations of the sources or targets can be estimated by measuring the spectral contents of the received signals.

In the biomedical engineering application, the spectral analysis of the signals from a patient provides doctors useful information for diagnosis purpose. In practice, the observed data are often of finite time-duration; hence, the quality of the spectral estimation is usually limited by the shortness of the available data record.

As a general rule for stationary random signals, the longer the data record, the better the spectral estimation that can be obtained. For deterministic signals, however, the spectral characteristics are described by an arbitrary length of data, the main goal being to select a data record as short as possible so that different signal components can be measured and resolved.

There are two broad classes of spectral analysis approaches: nonparametric methods and parametric (model-based) methods. The nonparametric methods—such as periodogram, Blackman-Tukey, and minimum variance spectral estimators—do not impose any model assumption on the data, other than wide-sense stationarity.

The parametric spectral estimation approaches, on the other hand, assume that the measurement data satisfy a generating model by which the spectral estimation problem is usually converted to that of determining the parameters of the assumed signal model. Two kinds of models are widely assumed and used within the parametric methods, according to different spectral characteristics of the signals: the rational transfer function (RTF) model and the sinusoidal signal model.

The RTF models, including autocorrelation (AR), moving average (MA), and autocorrelation moving average (ARMA) types are usually used to analyze the signals with continuous spectra, while the sinusoidal signal model is a good approximation of signals with discrete spectral patterns.

The discussion of PCG spectral analysis will divided into two parts: stationary spectral analysis and non-stationary spectral analysis.

In the first part, the nonparametric spectral estimation methods will introduced, along with the parametric methods for rational spectral analysis and sinusoidal spectral analysis.

In the second part, study of two non-stationary spectral analysis examples: damped sinusoidal parameter estimation, as approximation to the PCG-signal components, and instantaneous frequency measurement.

The typical spectral distribution of PCG signal can be well demonstrated in Fig. 5.1, as the indication of normal frequency PCG signal trace distribution, as well as the pathological PCG traces (pulmonary and mitral valve regurgitation) can be observed as spectral distribution as shown in Figs. 5.1 and 5.2.

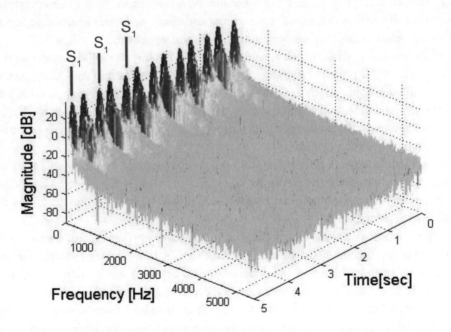

Figure 5.1: Normal eight PCG signal traces with corresponding spectral distribution over frequency range and through time course of signal recording.

This pathological condition related to the valvular diseases such as mitral valve regurgitation (MVR), which is considered in many PCG signal processing problems [34, 35]. The other cardiovascular pathologies also can be considered for spectral analysis and further post-processing steps. Some of investigators were focused on metabolic cardiac disorders such as diabetes mallietus (DM) and congestive heart disease (CHD), which shows some distinct characteristics in frequency domain as a priori- assumption for the spectral PCG-signal processing [35, 36].

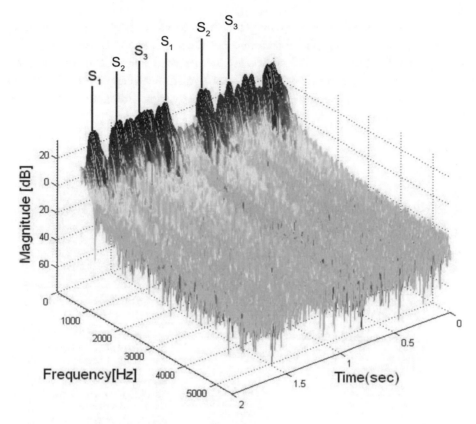

Figure 5.2: Mitral-valve regurgitation PCG-signal trace and its related spectral distribution over frequency range and corresponding time course of recording.

5.1.1   ENERGY SPECTRAL DENSITY OF DETERMINISTIC SIGNALS

$X(t)$ represents a continuous-time signal of interest; $x(n)$ denotes the sequence obtained by sampling x, suppose that x_C at some uniform sampling rate F_s; that is,

$$x(n) = x_c(n/F_s). \tag{5.1}$$

If $x(n)$ has finite energy, then the quantity $S(\omega)$ can be interpreted as the distribution of the signal energy as a function of frequency and, hence, it is called the energy spectral density of the signal. Here, the frequency is measured in radians per sampling interval, which corresponds to the physical frequency $F/2.F_s$ and corresponding unit in hertz. Note that the total energy of the signal is the integral of $S(\omega)$ over the interval (t,s) (within a constant scale 1/2).

If the autocorrelation function (ACF) of the deterministic signals (e.g., PCG signal) can be defined as x, therefore, the PCG signal energy index can be obtained from integration of ACF in frequency domain.

The flow diagram of the PCG signal energy detection based on spectral estimation is presented in Fig. 5.3, where three PCG microphone transducers were acquired and the PCG signals sampled at Shannon frequency F_s, then the PCG signals are band-passed filtered and the relevant spectral estimation was derived. Further higher-order statistics algorithm was applied to the PCG spectral patterns, such as a k-mean clustering method and the energy signature identification of the heart was detected and evaluated [74, 77].

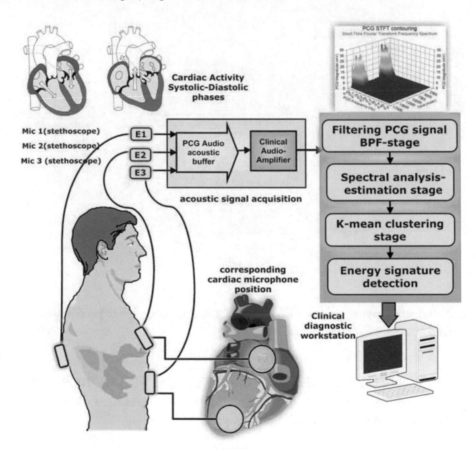

Figure 5.3: Phonocardiography energy detection flow diagram based on spectral estimation with higher-order clustering to identify the energy signature of the heart, in order to discriminate cardiac work during systolic-diastolic phase.

5.1.2   PCG SPECTRAL ESTIMATION

PCG spectra contain vast number of distinct harmonics categories that would be useful to be identified as a clustering scheme in any data classification algorithms. The majority of these spectra, although, belong to specific valvular pathologies which have a distinct energy (intensity) level in STFT plot. This realness would be an attractive point to consider such variation as clustering index, as this will considerably orientate the classifier algorithm to stable entropy value.

The spectral information characteristics of PCG lie within a frequency band of (54–520 Hz), and this band depend on digital stethoscopic interface and resolution of data converters in instrumentation platform. This criteria constitutes the basis for heart sounds spectral pattern classification technique, in which the dependency on frequency (spectral) characteristics.

The block diagram for overall spectral classification system is demonstrated in Fig. 5.3. Several patterns can be derived from the input-vector of PCG signal, in which they are processed with a specific FIR-filter length.

The most recognizable patterns in PCG are systolic, diastolic, pre-systolic, and post-diastolic peaks, of successive heart sound ($S_1, S_2, S_3,$ and S_4), which are shown in Fig. 5.1. Most of cardiologists prefer the base-diagnosis on two categories of PCG, S_1 and S_2, so that they can discriminate the hemodynamics turbulences (normal-level deviating blood flow) and cardiovascular anomalies in an appropriate method.

The spectra stamp can be oriented in three schema (supraspectra, infraspecta, and mid-spectra) which represent the intensity harmonics of PCG waveform. The correlation indicator between two intensity peaks of PCG, where it gives a defined index for clustering profile M_{j-PCG} of PCG signal, which in turn apply a segmental cluster for input data vector [79].

Systolic and diastolic murmur frequencies are classified by the frequency band containing the largest power value in the tenth (s) of the systole/diastole corresponding to the found maximum values of (SI/DI). If the largest power value is found in one of the two lowest frequency bands (containing frequencies below 125 Hz), the murmur is classified as a low-frequency murmur. If the largest power value is found in one of the eight highest frequency bands (containing frequencies above 250 Hz), the murmur is classified as a high-frequency murmur. If the none of the above is the case, the murmur is classified as a medium-frequency murmur [75, 77].

- The PCG spectral estimation: This result is obtained by using Db-wavelets decomposition techniques for a set of PCG signal as below:

$$y[n] = (x_{PCG} * g)[n] = \sum_{k=-\infty}^{\infty} x_{PCG}[k]g[n - k]. \tag{5.2}$$

Extracting the PCG diastolic low-frequency components as in Eq. (5.1):

$$y_{low PCG(diastolic)}[n] = \sum_{k=-\infty}^{\infty} x_{PCG}[k]g[2n - k]. \tag{5.3}$$

And for PCG systolic high-frequency components:

$$y_{highPCG(diastolic)}[n] = \sum_{k=-\infty}^{\infty} x_{PCG}[k]h[2n - k]. \tag{5.4}$$

Based on the spectral characteristic features extracted from the heart sound signal, the nature of the heart sound can be identified using pattern recognition techniques. A number of pattern recognition and classification schemes have been implemented for the analysis of heart sounds. The classical pattern recognition techniques, includes the Gaussian-Bayes classifier, the K-nearest neighbor classifier (k-mean clustering), and higher-order classification algorithms.

The Gaussian-Bayes classifier is the most popular parametric technique of supervised pattern recognition. It is considered optimal when the probability density functions (p.d.f) of the patterns in the feature space are known (a pattern is defined as an N-dimensional vector composed of (N) features), but it needs high computational efficiency and pre-definition of a priori conditions to increase the accuracy of this classifier type [79].

The K-nearest neighbor classifier is a nonparametric approach, which is useful when the probability density functions are difficult to estimate or cannot be estimated [80].

The nearest neighbor method is an intuitive approach based on distance measurements, motivated by the fact that patterns belonging to the same class should be close to each other in the feature space. Joo et al [80]. demonstrated the diagnostic potential of a Gaussian-Bayes classifier for detecting degenerated bioprostheses implanted in the aortic valve position. A detailed discussion will be presented in chapter six, with different PCG-classifiers and its relation with spectral estimation approach [80, 82].

5.2   NONPARAMETRIC METHOD FOR PHONOCARDIOGRAPHY SPECTRAL ESTIMATION

The spectral estimation methods are based on the discrete Fourier transform (DFT) of either the signal segment or its autocorrelation sequence. In contrast, parametric methods assume that the available signal segment has been generated by a specific parametric model (e.g., a pole-zero or harmonic model). Since the choice of an inappropriate signal model will lead to erroneous results, the successful application of parametric techniques, without sufficient a priori information, is very difficult in practice. If adopting a deterministic signal model, the mathematical tools for spectral analysis are the Fourier series, which is discussed in chapter four.

It should be stressed at this point that applying any of these tools requires that the signal values in the entire time interval from $-\infty$ to $+\infty$ be available. If it is known a priori that a signal is periodic, then only one period is needed. The rationale for defining and studying various spectra for deterministic signals is threefold. First, we note that every realization (or sample function) of a stochastic process is a deterministic function.

Therefore, we can use the Fourier series and transforms to compute a spectrum for stationary processes. Second, deterministic functions and sequences are used in many aspects of the study of stationary processes, for example, the autocorrelation sequence, which is a deterministic sequence. Third, the various spectra that can be defined for deterministic signals can be used to summarize important features of stationary processes.

Most practical applications of spectrum estimation involve continuous-time signals. For example, in speech analysis we use spectrum estimation to determine the pitch of the glottal excitation and the formants of the vocal tract [74, 75]. In electroencephalography (EEG), we use spectrum estimation to study sleep disorders and the effect of medication on the functioning of the brain [76]. Another application is in ultrasonic Doppler radar, where the frequency shift between the transmitted and the received waveform is used to determine the radial velocity of the target [83].

The main numerical computation of the spectrum of a continuous-time signal involves the following three steps.

1. Sampling the continuous-time signal to obtain a sequence of samples.

2. Collecting a finite number of contiguous samples (data segment or block), to use for the computation of the spectrum. This operation, which usually includes weighting of the signal samples, is known as windowing, or tapering.

3. Computing the values of the spectrum at the desired set of frequencies.

This last step, is usually implemented using some efficient implementation of the DFT. The above processing steps, which are necessary for DFT-based spectrum estimation, are shown in Fig. 5.4. The continuous-time signal is first processed through a low-pass (anti-aliasing) filter and then sampled to obtain a discrete-time signal. Data samples of frame length N with frame overlap N_0 are selected and then conditioned using a window. Finally, a suitable-length DFT of the windowed data is taken

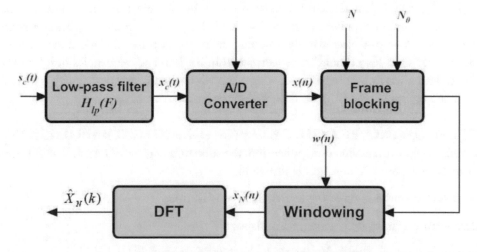

Figure 5.4: DFT-based Fourier analysis system for continuous-time signals.

as an estimate of its spectrum, which is then analyzed. In this section, we discuss in detail the effects of each of these operations on the accuracy of the computed spectrum. The understanding of the implications of these effects is very important in all practical applications of spectrum estimation.

The above processing steps, which are necessary for DFT-based spectrum estimation, are illustrated in Fig. 5.4. The continuous-time signal is first processed through a low-pass (anti-aliasing) filter and then sampled to obtain a discrete-time signal. Data samples of frame length N with frame overlap N_0 are selected and then conditioned using a window.

Finally, a suitable-length DFT of the windowed data is taken as an estimate of its spectrum, which is then analyzed. In this section, we discuss in detail the effects of each of these operations on the accuracy of the computed spectrum. The understanding of the implications of these effects is very important in all practical applications of spectrum estimation.

5.2.1   EFFECT OF SIGNAL SAMPLING

The continuous-time signal $s_c(t)$, whose spectrum we seek to estimate, is first passed through a low-pass filter, also known as an anti-aliasing filter $H_{lp}(F)$, in order to minimize the aliasing error after sampling. The antialiased signal $x_c(t)$ is then sampled through an analog-to-digital converter (ADC) to generate the discrete-time sequence x(n), that is,

$$x(n) = x_c(t)\,\big|_{t=n/Fs}.\tag{5.5}$$

From the sampling theorem which introduced before

$$X(e^{j2\pi F/F_s}) = F_s \sum_{l=-\infty}^{\infty} X_c(F - lF_s),\tag{5.6}$$

where $X_c(F){=}H_{lp}(F)S_c(F)$. notice that the spectrum of the discrete-time signal x(n) is a periodic replication of $X_c(F)$. Overlapping of the replicas $X_c(F\text{-}l\,F_s)$ results in aliasing. Since any practical anti-aliasing filter does not have infinite attenuation in the stop-band, some nonzero overlap of frequencies higher than $F_{s/2}$ should be expected within the band of frequencies of interest in x(n). These aliased frequencies give rise to the aliasing error, which, in any practical signal, is unavoidable. It can be made negligible by a properly designed anti-aliasing filter H_{lp}.

5.2.2   WINDOWING, PERIODIC EXTENSION, AND EXTRAPOLATION

In practical signal spectral estimation application, the spectrum of a signal by using a finite-duration segment can be computed. The reason is threefold:

1. The spectral composition of the signal changes with time.

2. There is only a finite set of data at this disposal.

3. The computational complexity should be kept to an acceptable level.

Therefore, it is necessary to partition x(n) into blocks (or frames) of data prior to processing. This operation is called frame blocking, and it is characterized by two parameters: the length of frame

N and the overlap between frames N_0 (see Fig. 5.4). Therefore, the central problem in practical frequency analysis can be stated as follows:

Determine the spectrum of a signal x(n) $(-\infty < n < \infty)$, from its values in a finite interval $(0 \leq n \leq N\text{-}1)$, that is, from a finite-duration segment.

Since x(n) is unknown for n < 0 and n $\geq$ N, it is difficult to say, without having sufficient a priori information, whether the signal is periodic or aperiodic. If we can reasonably assume that the signal is periodic with fundamental period N, we can easily determine its spectrum by computing its Fourier series, using the DFT method.

However, in most practical applications, This assumption can not be considered, because the available data block could be either part of the signal period of a periodic function or a segment from an aperiodic function. In such cases, the spectrum of the signal cannot be determined without assigning values to the signal samples outside the available interval. There are three methods to deal with this issue:

1. Periodic extension method. This method is based on assumption that x(n) is periodic with period N, that is, x(n)=x(n+N) for all n and the Fourier series coefficients can be computed, using the DFT-method.

2. Windowing-method. This process can be initialized by supposing that the signal is zero outside the interval of observation, that is, x(n)=0 for n < 0 and n $\geq$ N. This is equivalent to multiplying the signal with the rectangular window.

3. Extrapolation method. This process uses a priori information (pre-assumption) about the signal to extrapolate (i.e., determine its values for n < 0 and n $\geq$ N) outside the available interval and then determine its spectrum by using the DTFT.

The periodic extension and windowing can be considered the simplest forms of extrapolation. It should be obvious that a successful extrapolation results in better spectrum estimates than periodic extension or windowing. Periodic extension is a straightforward application of the DFT, whereas extrapolation requires some form of a sophisticated signal model, as previously introduced of the nonparametric spectral estimation methods. Therefore, we can considering the PCG-spectral estimation technique, based on non-parametric method.

Firstly, the PCG-periodogram estimator will introduced, and analysis of its statistical properties in terms of the bias and the variance of the PSD estimate will discussed. Since the periodogram estimator has a high variance, even for large sample length, several modified methods such as Bartlet [71], Welch [73], and Blackman–Tukey [74] methods are then discussed. Finally, the minimum variance spectral estimator is given.

5.2.3 PERIODOGRAM METHOD

The periodogram spectral estimator of a phonocardiography (PCG) is defined based on the Power Spectral Density (PSD) equation as follows:

$$\hat{R}_{PCG}(\omega) = \frac{1}{N} \left| \sum_{n=0}^{N-1} x(n)e^{-j\omega n} \right|^2 = \frac{1}{N} |X(\omega)|^2, \tag{5.7}$$

where $\hat{R}_{PCG}(\omega)$ is the periodogram of PCG-spectral estimation, and the PSD equation can be represented as follows:

$$R_{PCG}(\omega) = \lim_{x \to 0} E \left[\frac{1}{N} \left| \sum_{n=0}^{N-1} x(n)e^{-j\omega n} \right|^2 \right], \tag{5.8}$$

where $X(\omega)$ is the Fourier transform of the sample sequence x(n).

Note that the implementation of the periodogram estimator involves performing discrete Fourier transform (DFT) on x(n), followed by calculating the PSD directly. Specifically, given (N) data points x(0), x(1),..., x(N-1), we compute the N-point DFT at frequency.

$$\omega = \frac{2\pi}{N}k, \text{ k} = 0,1,...,\text{N-1}. \tag{5.9}$$

that yields the samples of the periodogram

$$\hat{R}_1(\frac{2\pi}{N}k) = \frac{1}{N} \left| \sum_{n=0}^{N-1} x(n)e^{-j2\pi n \frac{k}{N}} \right|^2, \text{ k} = 0,1,...,\text{N-1}. \tag{5.10}$$

In practice, however, when the data length N is small, the estimated PSD computed by Eq. (5.8) does not provide a good representation of the continuous spectrum estimate due to the small number of samples. In order to get a more complete description about the estimated PSD, it is necessary to evaluate $\hat{R}_{PCG}(\omega)$ at more dense PCG frequencies. This can be achieved by increasing the sequence length via zero padding. Specifically, if the data length is increased to L (L>N), evaluating L-point DFT yields the following:

$$\hat{R}_2(\frac{2\pi}{L}k) = \frac{1}{N} \left| \sum_{n=0}^{N-1} x(n)e^{-j2\pi n \frac{k}{L}} \right|^2, \text{ k} = 0,1,...,\text{L-1}. \tag{5.11}$$

Figure 5.5 illustrates the correlation matrix of PCG signal after applying the periodogram evaluation process described above. This figure plotting the correlation-index derived from Eq. (5.11) against PCG intensity amplitude, to identify the spectral properties of phonocardiography traces.

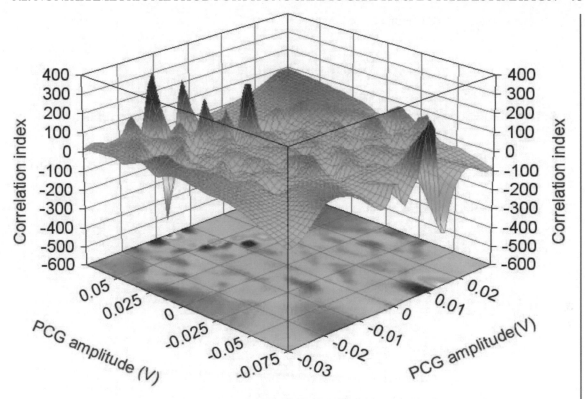

Figure 5.5: Correlation 3D-contouring of PCG signal based on periodogram method

5.2.4   MODIFIED PERIODOGRAM METHOD

The Bartlett method [78] and the Welch method [79] are two modified periodogram methods. These methods aim at reducing the variance of the periodogram estimate by splitting up the N available observations into K segments, and then averaging the periodograms computed from each segments for each value of ω. Therefore let us consider the following:

$$x_i(n) = x(n + iD), i = 0, 1, ..., K - 1; \text{n=0, 1,...,M-1}. \tag{5.12}$$

Denote the observations of the i_{th} segment, where (iD) is the starting point of the i_{th} segment. The Bartlett method takes D=M, and N=L.M; thus, data samples in successive segments are not overlapped. In the Welch method, one chooses D=M/2 and obtains overlapped data samples in successive segments. For example, if D=M/2, there is 50% overlapping between successive data segments, and K=2L segments are obtained.

Let's assume the equations below.

$$\hat{R}^i(\omega) = \frac{1}{M}\left|\sum_{n=0}^{M-1} x_i(n)e^{-j\omega}\right|^2 \tag{5.13}$$

represents the periodogram of the ith segment. The Bartlett spectral estimator is defined as

$$\hat{R}_B(\omega) = \frac{1}{L}\sum_{i=0}^{L-1}\hat{R}^i(\omega). \tag{5.14}$$

The Welch spectral estimator is defined as:

$$\hat{R}_W(\omega) = \frac{1}{K}\sum_{i=0}^{K-1}\hat{R}_M^i(\omega), \tag{5.15}$$

where $R_M^i(\omega)$ is the window-based periodogram, given by:

$$\hat{R}_M^i(\omega) = \frac{1}{MP}\left|\sum_{n=0}^{M-1} x_i\omega(n)e^{-j\omega}\right|^2 \tag{5.16}$$

with P the (power) of the PCG-signal time window w(n),

$$P = \frac{1}{K}\sum_{n=0}^{M-1}\omega^2(n). \tag{5.17}$$

It is noted that in the Welch method, the data samples in each segment are windowed before they are Fourier transformed via FFT-method. The statistical properties of the Bartlett estimator are easily obtained. First, the expected value of $\hat{R}_B(\omega)$ is given by:

$$E\left[\hat{R}_B(\omega)\right] = \frac{1}{L}E\left[\hat{R}^i(\omega)\right] = \frac{1}{2\pi}\int_{-\pi}^{\pi} R(\alpha)W_B^M(\omega-\alpha)d\alpha \tag{5.18}$$

$$W_B^{(N)}(\omega) = \frac{1}{N}\left[\frac{sin(\omega 2N)}{sin(\omega 2N)}\right]^2 \tag{5.19}$$

which is the Fourier transform of the so-called Bartlett window with length N, described as follows:

$$W_B^N(k) = \begin{cases} 1 - \frac{|k|}{N}, if\ |k| \le N-1 \\ 0, \qquad otherwise \end{cases}. \tag{5.20}$$

5.3   PARAMETRIC METHOD FOR PHONOCARDIOGRAPHY SPECTRAL ESTIMATION

5.3.1   PCG ARMA SPECTRAL ESTIMATION

Adaptive system identification techniques have been applied to model and identify biomedical signals such as the ECG, and PCG. One objective of PCG signal analysis is to extract features of heart sound signal, which are useful for early detection of hemodynamics abnormalities. The modeling and simulation of PCG's tracing, which are recordings of heart sounds, have great interest for cardiologist.

The autoregressive moving average (ARMAX) method, adaptive ARMAX modeling method, is used to estimate acoustic dynamics parameters of PCG signals. The analysis yields spectral features for use in classifying patient PCG's as normal or abnormal pattern to be assist in clinical diagnosis. Several diagnostic methods are available for detection of malfunctioning heart valves (mitral, aortic, and tricuspid), or for different gallop-disorders (diastolic and systolic gallop, atrial, and ventricular gallop).

The existing PCG signal acquisition techniques can be separated into invasive techniques, which make measurements inside the body, and noninvasive, which operate external to the body. Noninvasive methods can be further divided into active methods, which transmit and receive a signal, and passive methods, which merely listen to signals generated by myocardium vibration [76, 79].

By assuming that, the PCG signal generated by passing a zero-mean white noise process u(n) through a linear time invariant (LTI) system; that is, accordingly, the definition of an ARMA signal is obtained by filtering a white noise process through a pole-zero system. The ARMA models are suitable for describing signals whose spectra have both sharp peaks and deep nulls by relatively lower orders where

$$x(n) = -\sum_{k=1}^{p} a(k)x(n-k) + \sum_{k=0}^{q} b(k)u(n-k),$$ (5.21)

where u(n) is called driving noise, and without loss of generality, b(0)=1. The corresponding system transfer function is as follows:

$$H(z) = \frac{B(z)}{A(z)},$$ (5.22)

where

$$A(z) = 1 + \sum_{k=1}^{p} a(k)z^{-k}$$ (5.23)

and

$$B(z) = \sum_{k=0}^{q} b(k)z^{-k}.$$ (5.24)

Therefore, the three rational transfer function model derived is as follows:

$$H(z) = \frac{1 + \sum_{k=1}^{p} a(k)z^{-k}}{\sum_{k=0}^{q} b(k)z^{-k}}.$$ (5.25)

- Autoregressive moving average (ARMA) model. The one pole-zero model in Eq. (5.25) is said to be an ARMA model of orders p and q and is denoted as ARMA (p,q). a(k) and b(k)'s (p and (q) are referred to as AR and MA coefficients orders, respectively.

- Autoregressive (AR) Model. If q=0, the model in Eq. (5.25) is simplified to an all-pole model with order p and is referred to as an AR(p)model.

- Moving average (MA) model. If p=0, the model in Eq. (5.25) is reduced to an all-zero model with order q, and is called MA(q) model.

The ARMA spectral estimation can be defined through auto-correlation function of ARMA transfer function model as follows:

$$r(m) = \left\{ -\sum_{k=1}^{p} a_k r(m-k) + \sigma^2 \sum_{k=1}^{q-m} h_k b(k+m), \text{m=0,1,...,q}; \ -\sum_{k=1}^{p} a_k r(m-k), m \geq q+1; \right.$$

$$(5.26)$$

Error biased value of estimated PCG signal was presented in Fig. 5.6, which is denoted as (erb-plot) of the stationary spectral behavior of the PCG-signals. Therefore, the ARMA parameters appear

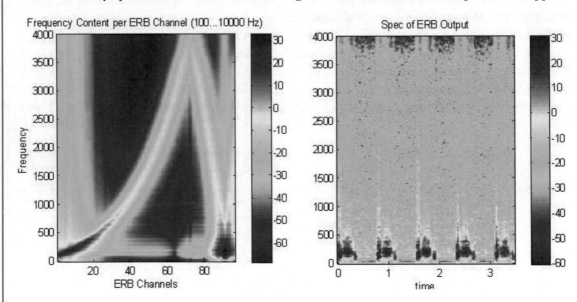

Figure 5.6: Spectral representation of cardiac PCG signal with assigned to aortic stenosis.

in a nonlinear fashion through the unknown impulse response h(n). If the optimum modeling is required, it is necessary to solve the least mean square solution of the highly nonlinear Yule-Walker equations. To obtain such a solution, nonlinear iterative techniques are employed, which are not only computationally expensive, but also suffer from the local convergence effect.

A considerable simplicity in computation may be achieved, via the suboptimal techniques in which the AR and MA part coefficients are estimated separately. At that point, it is possible to estimate the AR parameters via a linear procedure. After the AR parameters are obtained. The AR polynomial to filter the observed data will be used, and obtaining a pure MA(q) process, whose parameters can be estimated via the approaches developed in the preceding subsection.

AR Parameter Estimation. Choosing m≥q+1 in Eq. (5.27), and the following equation is obtained:

$$\sum_{k=0}^{p} a_k r(m-k), \text{ m=q+1,q+2,...,q+n.} \tag{5.27}$$

Equation (5.27) establishes a linear relation between the AR parameters and the ACFs of the observed signals. To determine the AR parameters, one may adopt the first p linear equations (i.e., q+1 ≤ m ≤ q+p) and then solve the resultant system of equations. When the ACFs are truly known, this set of equations is enough to yield a unique and accurate solution to the AR parameter estimates.

In practice, since the sample ACF estimates are used, the AR parameter estimates obtained by this method may be poor due to the estimation errors of the sample ACF estimates. This deficiency may also be interpreted by the fact that only subset lags of ACF's are used. In fact, Eq. (5.27) is satisfied for any m ≥ q+1.

To obtain better AR parameter estimates, one reasonable choice is to employ more than the minimal number (i.e., p) of the extended Yule-Walker equations. This results in an overdetermined set of linear equations which can be solved via least square (LS) or total least square (TLS) techniques. The adaptive filtering architecture, which is used in PCG signal spectral estimation, can be observed in Fig. 5.7 where the ARMA-PCG model based on auto correlation function used in define adapted filtering response in the presence of background noise signal n(t) or k[n] as displayed in Fig. 5.7.

5.3.2 ARMA MODELING APPROACH

For the clarity of statement, the real-valued signals should only be considered in developing the ARMA modeling approach for sinusoidal frequency estimation. The sinusoids in additive white noise satisfy a special ARMA model by which an ARMA modeling approach is developed for estimating the sinusoidal parameters was firstly, approved.

To motivate the selection of an ARMA process as the appropriate model for sinusoids in white noise, let us consider the following trigonometric identity:

$$cos(\Omega n) = -2.cos\Omega cos[\Omega(n-1)] - cos[\Omega(n-2)] \tag{5.28}$$

for $-\pi \le \Omega \le \pi$. Let x(n)=cosΩn, a(1)=2cosΩ, and a(2)=1; the single real sinusoidal component x(n) can be generated via the second-order difference equation:

$$x(n) = -a(1)x(n-1) - a(2)x(n-2) \tag{5.29}$$

with the initial values to be x(-1)=-1, x(-2)=0. This difference equation has the characteristics polynomial expression

$$1 + a(1)z^{-1} + a(2)z^{-2} \tag{5.30}$$

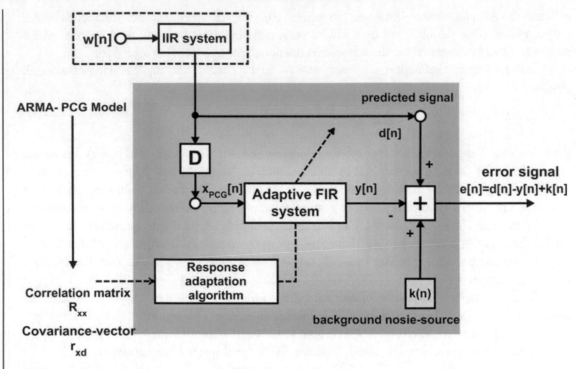

Figure 5.7: Schematics of phonocardiography simulation for adaptive FIR-model prediction process, where in the input PCG-signal x[n] to IIR-filtering; d[n] is the predicted signal, R_{xx},r_{xd} are the correlation matrix and covariance vector, respectively, the y[n] is the adaptive FIR-filtering system output; and k[n] is the background noise-source in the adaptive PCG-processing loop.

whose roots are $z_1 = e^{j\omega}$ and $z_2 = z^*1 = e^{j\Omega}$. The sinusoidal frequency is determined from the roots as follows:

$$\Omega = tan^{-1}(Imz_1/Rez_1). \tag{5.31}$$

The block diagram of the overall ARMA model system can be identified through the zero's and pole's transfer function estimation process as shown in Fig. 5.8. The other modified method for estimating PCG parametric model was based on adaptive regressive-parametric modeling, which will not be considered here because it belongs to the nonlinear processing approach of PCG.

5.3.3   PHONOCARDIOGRAPHY ESPRIT METHOD

ESPRIT (Estimation of Signal Parameters via Rotational Invariance Techniques [74, 75]) is another Eigen-decomposition method for estimating sinusoidal frequency parameters. It produce the sinusoidal frequency estimates by computing the generalized eigenvalues of two well-constructed matrices. We again, consider the complex-valued case.

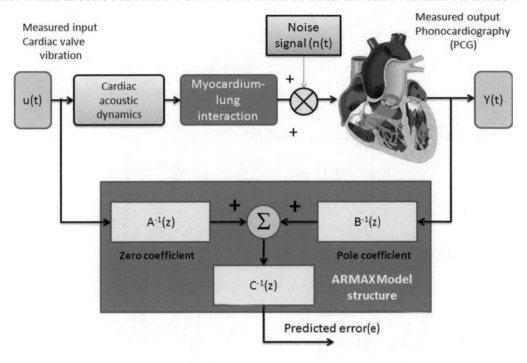

Figure 5.8: Block diagram of phonocardiography identification based on adaptive ARMA modeling technique. The measured input signal represents phonocardiograph time-function, successively this signal interact with cardiac acoustic dynamics. The noise signal n(t) (disturbance), will be summed with physical variable, the two channel (input and output) are connected to the ARMA-identification model to extract pole and zero- coefficients for cardiac acoustic model.

Using the notations defined in the MUSIC method, and denoting the following formula:

$$z(n) = [y(n + 1), y(n + 2), ..., y(n + m)]^{T}. \tag{5.32}$$

Therefore, along with Eqs. (5.26) and (5.27) the following can produce:

$$z(n) = A\Phi x(n) + w(n + 1)$$

where Φ is a $p \times p$ diagonal matrix.

$$\Phi = \text{diag}[e^{j\omega_1}, e^{j\omega_2}, e^{j\omega_p}] \tag{5.33}$$

which relates the time-displaced vector y(n) and z(n), and hence, is called a rotation operator. The cross-correlation matrix of the data vectors y(n) and z(n)is:

$$R_1 = E[y(n)z^{H}(n)] = APH\Phi A^{H} + \sigma^2 Q. \tag{5.34}$$

where, Q=identity matrix.

$$\begin{bmatrix} 1 & 0 & \dots & 0 & 0 \\ 0 & 1 & \dots & 0 & 0 \\ \vdots & \vdots & \ddots & \vdots & 0 \\ 0 & 0 & \dots & 1 & 0 \end{bmatrix} \overset{\triangle}{}$$

On the other hand, direct calculation of R_1 yields,

$$R_1 = \begin{bmatrix} r_y(1) & r_y(0) & \dots & r_y(m) \\ r_y(0) & r_y(1) & \dots & r_y(m-1) \\ \vdots & \vdots & \ddots & \vdots \\ r_y^*(m-2) & r_y^*(m-3) & \dots & r_y(1) \end{bmatrix} \tag{5.35}$$

and by constructing the following two matrices:

$$C_1 \overset{\triangle}{=} R - \sigma^2 I = APA^H \tag{5.36}$$
$$C_2 \overset{\triangle}{=} R_1 - \sigma^2 Q = AP\Phi^H A^H$$

and consider the matrix $(C_1 - \lambda C_2)$;

$$C_2 \lambda C_1 = AP(I - \lambda \Phi^H)A^H \tag{5.37}$$

Paularj, Roy, and Kailath [64, 65, 70] have shown that matrix pair (C_1,C_2) has (p) generalized eigenvalues at $\lambda(i)$. $e^{j\omega}$ i, i=1, 2, ..., p, and (m-p) generalized eigenvalues being zero.

Using the above results, the ESPRIT algorithm can be summarized as follows:

- Step 1. Calculating the sample ACF's $\hat{r}y(m)$, m=0,1,..., m., using a standard biased formula, and construct the matrices R and R_1 using Eqs. (5.35) and (5.36).

- Step 2. Computing the eigenvalues of R, and obtain the estimate of noise variance $\hat{\sigma}$.

- Step 3. Computing $\hat{C}_1.R ...2\hat{I}$ and $\hat{C}_1.R_1 ...2\hat{Q}$

- Step 4. Computing the generalized eigenvalues of the matrix pair $(\hat{C}_1, \hat{C}_2)$. The (p) generalized eigenvalues which lie on (or near) the unit circle determine the diagonal elements of , and hence, the sinusoidal frequencies. The remaining (m.p) eigenvalues will lie at (or near) the origin.

5.4   SPECTRAL-WINDOW METHOD FOR PCG-SIGNAL PROCESSING

Any signal in the world can be described in different coordinate systems and that there is engineering value in examining a signal described in an alternate basis-system. One basis system that is particularly useful is the set of complex exponentials.

The attraction of this basis set is that complex exponentials are the eigen-functions and eigen-series of linear time invariant (LTI) differential and difference operators, respectively. Put in its simplest form, this means that when a sine wave is applied to an LTI filter the steady-state system response is a scaled version of the same sine wave.

The system model can only affect the complex amplitude (magnitude and phase) of the sine wave but can never change its frequency. Consequently, complex sinusoids have become a standard tool to probe and describe LTI systems. The process of describing a signal as a summation of scaled sinusoids is standard Fourier transform analysis.

The Fourier transform and Fourier series, shown in Eq. (5.38), permits us to describe signals equally well in both the time domain and the frequency domain:

$$H(\omega) = \int_{-\inf}^{\inf} h(t)e^{-j\omega t} dt \tag{5.38}$$

$$H(\theta) = \int_{-\inf}^{\inf} h(n)e^{-j\theta n}, \tag{5.39}$$

$$h(t) = \frac{1}{2\pi} \int_{-\inf}^{+\inf} h(\omega)e^{+j\omega t} d\omega, \tag{5.40}$$

$$h(n) = \frac{1}{2\pi} \int_{-\pi}^{+\pi} h(\theta)e^{+j\theta n} d\theta. \tag{5.41}$$

Since the complex exponentials have infinite support, the limits of integration in the forward transform (time-to-frequency) are from minus to plus infinity. As observed earlier, all signals of engineering interest have finite support, which motivates us to modify the limits of integration of the Fourier transform to reflect this restriction. This is shown in Eq. (5.42), where T_{SUP} and (n) define the finite supports of the signal.

$$H_{SUP}(\omega) = \int_{T_{SUP}} h(t)e^{-j\omega t} dt \tag{5.42}$$

$$H_{SUP}(\theta)_{SUP} = \sum_{N} h(n)e^{-j\theta n}, \tag{5.43}$$

$$h(t) = \frac{1}{2\pi} \int_{-\inf}^{+\inf} H_{SUP}e^{+j\omega t} d\omega, \tag{5.44}$$

$$h(n) = \frac{1}{2\pi} \int_{-\pi}^{+\pi} H_{SUP}e^{+j\theta n} d\theta, \tag{5.45}$$

The two versions of the transform can be merged in a single compact form, if we use a finite support window to limit the signal to the appropriate finite support interval, as opposed to using the limits of integration or limits of summation.

5.5   DIGITAL STETHOSCOPE SYSTEM (DS)

The main elements of a Digital Stethoscope are the sensor unit that captures the heart and lung sounds (also known as auscultations), digitization, and digital processing of the auscultations for noise reduction, filtering and amplification. Sort of intelligent algorithms for heart rate detection and cardiac abnormalities detection may also be integrated in the DS-module.

Power source and battery management are key in this ultra-portable medical diagnostic tools, where key design considerations are ultra-low power consumption and high efficiency driven by the need for extended battery life, and high precision with a fast response time allowing quick determination of the patient's health status.

Additional requirements may drive the need for recording the auscultations, cabled, or wireless interfaces for transmission of the auscultations.

In addition, to enable ease of use, features like touch screen control and display back lighting are important for the device usability. Adding all these features without significantly increasing power consumption is a huge challenge. The selection of micro-processors units, instrumentation and buffer amplifiers, power sources and voltage regulators, Audio Codecs (ACOD) system, and both wired and wireless interface devices and ports provides the ideal tool box for digital stethoscope applications [80].

The main components of the cutting edge digital stethoscope and other clinical data fusion are shown in figure below where the use of low-power micro controller unit (MCU) in a portable medical instrumentation will assist in discretization and integration of multi-parametric cardiac diagnosis system.

The common core subsystems of a digital stethoscope are the following:

- Analog Front-End/Sensor Interface and Codec auscultations signal input is amplified and then digitized by the Audio Codec. Auscultations signal after being digitized and subjected to signal processing, is converted to analog and sent to the stethoscope earpieces.

- Low power micro-processor unit: Processors capable of executing all of the digital stethoscopes signal processing including key functions such as noise reduction, algorithms for heart rate detection, and heart defect detection while maintaining a very low constant current draw from the battery are good fit. The ability to control interfacing with memory and peripheral devices will be helpful.

  Given the nature of the device, processors that can manage the digital display and keyed functions allowing auscultation waveforms to be displayed and manipulated without additional components are ideal.

- Data Storage and Transmission: The auscultations can be recorded on MMC/SD card, or on a USB device. It can also be transmitted via wireless capability such as Bluetooth® and wireless transmission communication protocols such as IEEE 802.14 communication standards.

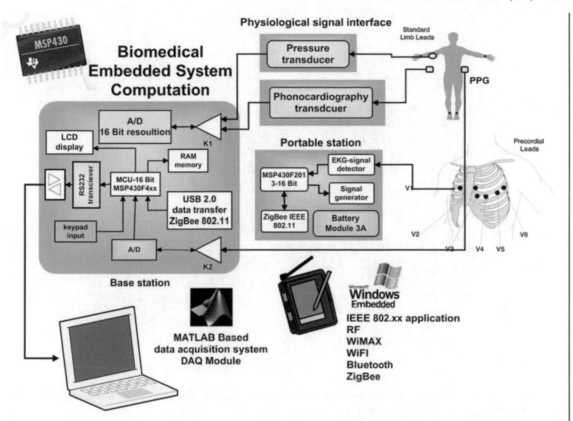

Figure 5.9: Embedded system component of digital stethoscope and ECG signal into one compact CDS-unit which will be used widely in health care facilities and clinical mobile system.

5.5.1   VISUAL ELECTRONIC STETHOSCOPE

Another variation of the digital stethoscope is the visual electronic stethoscope system (VES) which is considered a cutting-edge technology in modern auscultation instrumentation. The basic configuration of the multifunction stethoscope unit which mean it is capable of recording analysis and output preliminary diagnostic index for a clinical cardiac test paradigm.

The prototype of this device is shown in Fig. 5.10, which is developed by Contec Medical System Co. Ltd. CMS, Hukun, China. The first version of this stethoscope was based on the simple implementation of FFT in the signal processing chain with single channel buffering unit.

The direct display of PCG waveform on LCD, accompanied with other cardiac physiological parameters, is one of the attractive points in design. Further development and research toward a front-end cardiac visual stethoscope to map the four-auscultation sites on the thorax.

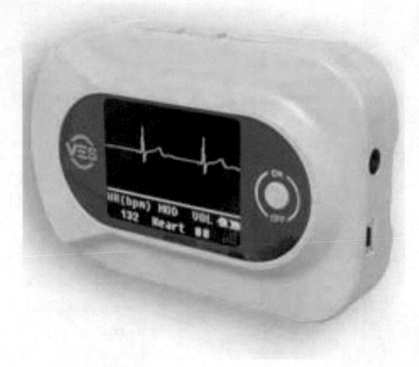

Figure 5.10: Visual electronic stethoscope (VES) system with multi-parametric physiological signals (SpO2, ECG, and heart sounds); Ref: CMS-VE, China.

5.6 SUMMARY

The main points of this chapter are as follows.

1. The spectral PCG signal analysis is an energetic tool for inspection of various cardiovascular disorders and their characteristics associated with variation in spectral attributes.

2. The two categories of spectral signal processing parametric and non-parametric, show interesting uses for detection of early and late cardiac diastolic index for several pathological conditions.

3. All the heart sound components, i.e., S_1, S_2, S_3, murmurs, and sound splits, were clearly separated in time and frequency. High resolution of generated heart sound images, in both time and pitch, are demonstrated, presenting a distinctly improved quality to classical spectrogram images (based on SFFT).

   The resulting visual images have self-referencing quality, whereby particular features and their changes become obvious at once.

4. characterization of heart sounds and uses both visual images derived from spectral plot and a system of integral plots that characterize time averaged and instantaneous sound intensity and frequency variations.

5. The digital stethoscope utilizes several spectral PCG signal processing in order to enhance its performance and ability to immediate diagnosis several hemodynamic disorders.

CHAPTER 6

PCG Pattern Classification

6.1 INTRODUCTION

PCG pattern classification, also known as auscultation pattern recognition, was one of the efficient computer-based methods applied to a medical decision-making system. The PCG classification, naturally, is based upon heart sound features. These PCG features can be represented as a set of electrical measurement of cardiac acoustic observations.

These measurements can also be represented in vector notation. Data features may also result from applying a feature extraction algorithm or operator to the input PCG data set. Significant computational effort may be required in feature extraction and the extracted features may contain errors or noise. Features may be represented by continuous, discrete, or discrete-binary variables.

Figure 6.1 presents the substantial block diagram of phonocardiography pattern classification method, the input of the system is the physiological signal, which represented here is the phonocardiography acoustic signal $(X_{pcg}(t))$ as a function of time. While the output of pattern recognition system could be statistical or syntactic representation, it depends on which computational method is used for feature extraction.

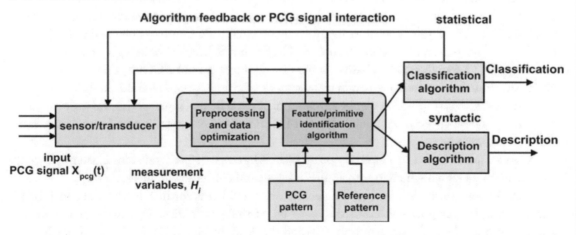

Figure 6.1: Substantial PCG pattern recognition system elements and signal flow diagram, which represents the principal strategy for PCG signal classification.

Ordinarily, the PCG features can be low-level patterns and high-level patterns. An example of PCG low-level patterns are signal intensities, as acquired from medical data acquisition as electrical signal, or as recorded as vibration signal. Examples of high-level patterns are vibration spectrum, PCG signal phase-delay, and energy profile.

Generally, PCG Pattern recognition is interpreted in two ways. The most general definition includes recognition of patterns in any type of PCG dataset and is called uniform PCG pattern classification. It discriminates peaks of heart sounds as excitation source for circulation hemodynamic, and is also called adaptive pattern clustering which magnifies and observes the spectral characteristics associated with PCG waveform turbulences and differentiate between them as clinical diagnostic indices or auscultation features.

6.2   PCG PATTERN CLASSIFICATION METHODS

PCG spectra contain a vast number of definite harmonics categories that would be useful in identifying clustering scheme in any data classification algorithms. The majority of these spectra belong to specific valvular pathologies having a distinct energy (intensity) level in STFT plot. This would be an attractive point to consider such variation as clustering point.

Considerably, this will orientate the classifier algorithm to a stable entropy value. The spectral information characteristics of PCG lie within a frequency band of 54-520 Hz and this band depends on digital stethoscopic interface and resolution of data converters in instrumentation platform.

This criteria is considered as an essential basis for PCG pattern classification technique in which the dependency on frequency (spectral) characteristics can be identified.

A block diagram for overall spectral classification system is demonstrated in Fig. 6.2 [81], where several patterns can be derived from the vector input PCG signal, in which they process with a specific FIR-filter length. The most recognized patterns in PCG are systolic and diastolic and presystolic and post-diastolic peaks of sound (S_1, S_2, S_3, and S_4). Most cardiologists prefer the base diagnosis of PCG signal, which is based on the two distinct peaks of PCG: S_1 and S_2. Therefore, they can discriminate the hemodynamics turbulences in the appropriate method. Additionally, they can sue the spectra stamp which can be oriented in three schema (supraspectra, infraspectra, and mid-spectra). These spectral segments represent the intensity harmonics for PCG waveform over defined time interval.

Correlation between two intensity peaks of PCG gives a defined index for clustering profile M_{j-PCG} of PCG signal which in turn applies a segmental cluster for input vector [81].

Systolic and diastolic murmur frequencies are classified according to the frequency band containing the largest power value in the tenth (s) of the systole/diastole, corresponding to the found maximum values of systolic interval (SI) and the diastolic interval (DI) (SI/DI). If the largest power value is found in one of the two lowest frequency bands (containing frequencies below 125 Hz), the murmur is classified as a low-frequency murmur. If the largest power value is found in one of the eight highest frequency bands (containing frequencies above 250 Hz), the murmur is classified as a high-frequency murmur. If none of the above is the case, the murmur is classified as a medium-frequency murmur [45].

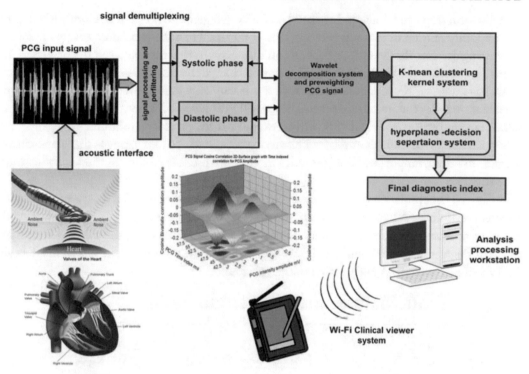

Figure 6.2: PCG signal pattern classification method as integrated in clinical-aided diagnosis system; during the first stage, the separation of systolic and diastolic component by aiding a wavelet-decomposition method, with definition of hyperplane using K-mean clustering technique. The complete analysis can be performed on medical console and the data can be transmitted to remote personal digital assistance (PDA) terminal through a wireless (Wi-Fi) communication protocol.

6.3   K-MEANS CLUSTERING METHOD

K-means clustering [66] differs in two important aspects from hierarchical clustering. Firstly, the K-means clustering requires the number of clusters-k beforehand. Secondly, it is not a hierarchical-based method; instead, it partitions the data set into (K) disjoint subsets. Again, the clustering is basically determined by the distances between objects. The K-means algorithm has the following structure: Algorithm applied to the PCG data vectors: K-means clustering approach;

1. Assign each object randomly to one of the clusters k=1,. . .,K.

2. Compute the means of each of the clusters.

3. Reassign each object to the cluster with the closest mean μk.

4. Return to step 2 until the means of the clusters do not change anymore.

The initialization step can be adapted to speed up the convergence. Instead of randomly labeling the data, K randomly chosen objects are taken as cluster means. Then the procedure enters the loop in step 3. Note again that the procedure depends on distances, in this case between the objects z_i and the means μ_k. Scaling the feature space will here also change the final clustering result.

An advantage of K-means clustering is that it is very easy to implement. On the other hand, it is unstable: running the procedure several times will give several different results. Depending on the random initialization, the algorithm will converge to different (local) minima.

In particular, when a high number of clusters is requested, it often happens that some clusters do not gain sufficient support and are ignored. The effective number of clusters then becomes much less than K.

In Fig. 6.3 the result of a K-means clustering is shown for a simple 2D data set representation. The means are indicated by the circles. At the start of the optimization, i.e., at the start of the trajectory, each mean coincides with a data object. After 10 iteration steps, the solution converged. The result of the last iteration is indicated by (x). In this case, the number of clusters in the data and the predefined K=3 match. A fairly stable solution is found.

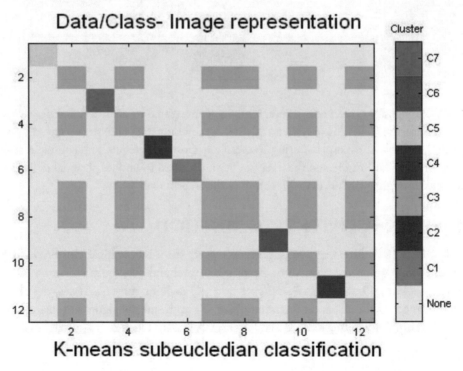

Figure 6.3: Sub-Euclidean K-mean classification result for seven input PCG signal ranging from aortic stenosis (AS), mitral regurgitation (MR), diastolic murmurs, and pulmonary valve stenosis (PS).

6.4   FUZZY C-MEANS CLASSIFICATION ALGORITHM

Fuzzy systems are currently finding practical applications, ranging from (soft) regulatory control in a variety of consumer (e.g., automotive, electro-mechanical, air-vacuum machine products, to accurate modeling of non-linear systems such as missile-guidance system, patient-ventilation device, cardiac pacemaker system, insulin-pumping unit [87]. Recently, a vast number of research focusing on analysis and pattern classification of biosignals based on fuzzy clustering methods; multi-resolution signal analysis and hybrid neuro-fuzzy network.

The development of the fuzzy c-means algorithm (FCM) [86, 88] was the birth of all clustering techniques corresponding to probabilistic clustering algorithm. The first version developed by Duda and Hart [80], by which they performed a hard cluster partition corresponding to definition of data cluster partition (hard c-means or hard ISODATA algorithm). In order to treat data belonging to several clusters to the same extent in an appropriate manner, Dunn [87] introduced a fuzzy version of this algorithm.

It was generalized once more in the final version by Bezdek [88] and his introduction of the data fuzzifier module. The resulting fuzzy c-means algorithm recognizes spherical clouds of points in a p-dimensional space, which represents the phonocardiography data channels from at least a two-channel description.

The clusters are assumed to be of approximately the same size. Each cluster is represented by its center. This representation of a cluster is also called a prototype, since it is often regarded as a representative of all data assigned to the cluster. As a measure for the distance, the Euclidean distance between a datum and a prototype is used.

The c-mean fuzzy clustering algorithm, is discussed here, was based on objective functions (J), which are mathematical criteria that quantify the goodness of cluster models that are comprised of prototypes and data partition.

Objective functions serve as cost functions that have to be minimized to obtain optimal cluster solutions. Thus, for each of the following cluster models the respective objective function expresses desired properties of what should be regarded as (best) results of the cluster algorithm.

In fuzzy clustering approach, the PCG data clustering have to be parameterized into four distinct criteria: intensity, frequency, spectral variation, and phase-shift. These criteria allow the partitioning of different clusters into an optimal cluster solution and for more than two classification of heart sounds specifications.

Having defined such a criterion of optimality, the clustering task can be formulated as a function optimization problem. That is, the algorithms determine the best decomposition of a dataset into a predefined number of clusters by minimizing their objective function. The steps of the algorithms follow from the optimization scheme that they apply to approach the optimum of (J).

Thus, in this presentation of the hard, fuzzy, and possibilistic c-means the respective objective functions first were discussed. In fuzzy clustering, each point has a degree of belonging to clusters, as in fuzzy logic, rather than belonging completely to just one cluster. Thus, points on the edge of a cluster, may be in the cluster to a lesser degree than points in the center of cluster. For each point

(x) there is a coefficient giving the degree of being in the k_{th} cluster $u_{k-PCGsignal}(x)$. Usually, the sum of those coefficients is defined to be unity-value:

$$\forall x \sum_{k=1}^{N} u_{k(PCG-signal)}(x) = 1 \tag{6.1}$$

according to the fuzzy c-means approach, the centroid of a PCG-data vector cluster is the mean of all data points, weighted by their degree of likelihood to the cluster (N). For the definition of cluster centroid, which is the principal step in c-mean classification technique:

$$C_k = \frac{\sum_{x-PCG} u_k(x)^m x}{\sum_{x-PCG} u_k(x)^m}. \tag{6.2}$$

Then the degree of likelihood (similarity) is related to the inverse of the distance to the data cluster center

$$u_{k-PCG}(x) = \frac{1}{d(C_{k-PCG}, x)}, \tag{6.3}$$

hence the coefficients are normalized and fuzzified with a real parameter (m = 1), so that their sum is (1)

$$u_{k-PCG}(x) = \frac{1}{\sum_j \left[\frac{d(C_{k-PCG}, x)}{d(C_{k-PCG}, x)}\right]^{2/(m-1)}}. \tag{6.4}$$

For m=2, this is equivalent to normalizing the coefficient linearly to make their sum 1. When m is close to 1, then the cluster center closest to the point is given much more weight than the others, and the algorithm is similar to k-means algorithm.

6.5   PRINCIPAL COMPONENT ANALYSIS (PCA)

The main purpose of principal component analysis (PCA) is to reduce the dataset dimensionality from (p) to (d), where d < p, while at the same time accounting for as much of the variation in the original data set as possible. With PCA, we transform the data to a new set of coordinates or variables that are a linear combination of the original variables. In addition, the observations in the new principal component space are uncorrelated. The hope is that this could gain information and understanding of the data by looking at the observations in the new space.

PCA is mathematically defined [85] as an orthogonal linear transformation that transforms the data to a new coordinate system such that the greatest variance by any projection of the data comes to lie on the first coordinate (called the first principal component), the second greatest variance on the second coordinate, and so on. PCA is theoretically the optimum transform for a given data in least square terms.

PCA can be used for dimensionality reduction in many biomedical and clinical data sets by retaining those characteristics of the data set that contribute most to its variance, by keeping lower-order principal components, and by ignoring higher-order ones. Such low-order components often contain the (most important) aspects of the data. However, depending on the application this may not always be the case.

The complete block diagram of principal component analysis technique (PCA) for the phonocardiography signal (PCG) is shown in Fig. 6.4, as a assisted clinical data classification module, which implemented in the ICU-online patient monitoring system. The application of PCA-analysis in

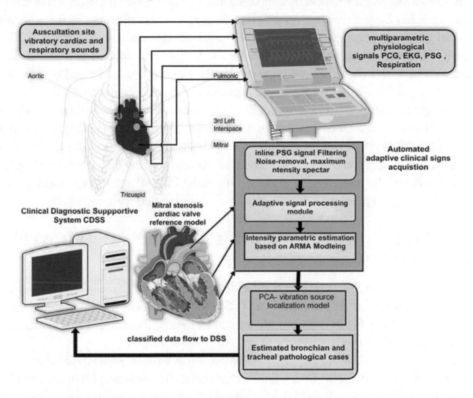

Figure 6.4: Block diagram of principal component analysis (PCA) of phonocardiography signal. The signal flow diagram, including four cardia microphone acquisition, directed to ICU-bedside monitoring. By applied in line-signal filtering and noise-attenuation module, the vibration sources from the 4-auscultation site were localized. In addition, this technique can be used for diagnosis of a respiratory disease by using 'lung sound analysis module' and tracheal-microphone to classify and estimate bronchial and tracheal pathological cases.

phonocardiography pattern classification was first used by Kuenter et al. (2002) where they combined methods for classification of high-dimension PCG data acquired from different patient groups for evaluating the performance of implanted heart valves.

For a PCG-data matrix, X_{PCG}^T, with zero empirical mean (the empirical mean of the distribution has been subtracted from the data set in previous filtering stage), where each row represents a different repetition of the auscultation experiment, and each column gives the results from a particular probe, the PCA transformation is given by:

$$Y^T = X_{PCG}^T W = V\Sigma, \qquad (6.5)$$

where $V\Sigma\, W^T$ is the singular value decomposition (SVD) of X^T.

PCA algorithm has the distinction of being the optimal linear transformation for keeping the subspace that has largest variance value. This advantage, however, comes at the cost of greater computational efficiency if compared, for example, to the discrete cosine transform (DCT).

Figure 6.5 shows the performance results for PCA algorithm applied to phonocardiography signals data vectors, where the above two plots display the signal inference index (SIR) for the seven PCG set indicating a robustness in discrimination of temporal variation in PCG signal.

The straightforward implementation of PCA-algorithm in different computational platform makes it attractive for different biomedical and biosignals higher-order classification and pattern recognition application. Moreover, the ability of PCA algorithm to reduce the clinical data high dimensionality, which may consume a vast amount of memory during continuous/real-time monitoring and recorded of the physiological signals. This make PCA-algorithm a favorable choice, for medical data abstraction system and in prototyping clinical decision making system.

6.6   HIGHER-ORDER STATISTICS PCG CLASSIFICATION PCG-HOS

Higher-order statistics (HOS) measures are extensions of second-order measures (such as the autocorrelation function (ACF) and power spectrum (PS)) to higher orders. The second-order measures work fine if the signal has a Gaussian (normal) probability density function, but as mentioned above, many real-life signals are non-Gaussian.

The easiest way to introduce the HOS measures is just to show some definitions, so that the reader can see how they are related to the familiar second-order measures. Here are definitions for the time-domain and frequency-domain third-order HOS measures, assuming a zero-mean, discrete signal x(n).

Time domain measures:

In the time domain, the second-order measure is the autocorrelation function,

$$R(m) = \langle x(n)x(n+m) \rangle, \qquad (6.6)$$

where $<>$ is the expectation statistics operator. The third-order measure is called the third-order moment

$$M(m1, m2) = \langle x(n), x(n+m1), x(n+m2) \rangle. \qquad (6.7)$$

Note that the third-order moment depends on two independent lags, m1 and m2. Higher-order moments can be formed in a similar way by adding lag terms to the above equation. The signal cumulants can be easily derived from the moments.

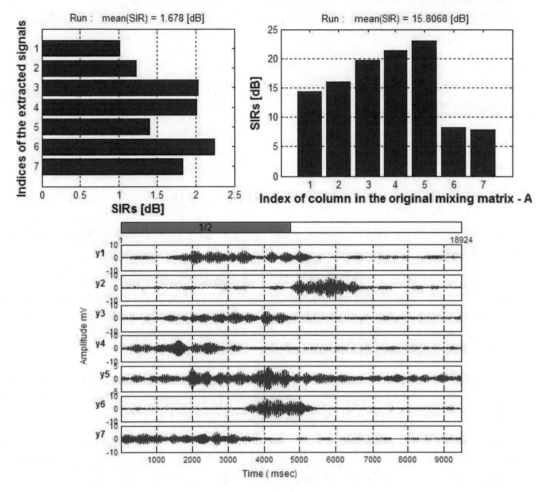

Figure 6.5: PCA decomposition of PCG signal, where the above-left graph represents the extracted 7-PCG signal trace SIR index with the mean of 1.678 dB, and the above-right graph represents the energy level of the same PCG trace in the original mixing PCA-scheme, with a mean SIR value of 15.8068 dB. The bottom graph represents decomposed PCG trace with PCA-algorithm, according to Eq. 6.5.

Frequency domain measures:

In the frequency domain, the second-order measure is called the power spectrum $P(k)$, and it can be calculated in two ways:

- Taking a Discrete Fourier Transform (DFT) of the autocorrelation function $R(m)$; $P(k)=DFT[R(m)]$;

or:

- Multiplying together the signal Fourier Transform X(k) with its complex conjugate; P(k)=X(k) X*(k)

At third-order the bispectrum B(k,l) can be calculated in a similar way:

- Taking a Double Discrete Fourier Transform (DDFT) of the third-order cumulant; B(k,l)=DDFT[M(m1,m2)];

or:

- Form a product of Fourier Transforms at different frequencies:

$$B(k, l) = X(k)X(l)X * (k + l). \tag{6.8}$$

The main parameters to be consider when dealing with (HOS) pattern classification which can summarized as below:

- Moments: Moments are statistical measures which characterize signal properties. We are used to using the mean and variance (the first and second moments, respectively) to characterize a signal's probability distribution, but unless the signal is Gaussian (Normal) then moments of higher orders are needed to fully describe the distribution. In practice, in HOS we usually use the cumulants rather than the moments.

- Cumulants: The n_{th} order cumulant is a function of the moments of orders up to (and including) (n). For reasons of mathematical convenience, HOS equations/discussions most often deal with a signal's cumulants rather than the signal's moments.

- Polyspectra: This term is used to describe the family of all frequency-domain spectra, including the second order. Most HOS work on polyspectra focusing attention on the bispectrum (third-order polyspectrum) and the trispectrum (fourth-order polyspectrum).

- Bicoherence: This is used to denote a normalized version of the bispectrum. The bicoherence takes values bounded between 0 and 1, which make it a convenient measure for quantifying the extent of phase coupling in a signal. The normalization arises because of variance problems of the bispectral estimators for which there is insufficient space to explain.

6.7   INDEPENDENT COMPONENT ANALYSIS (ICA) METHOD

The ICA approach and blind signal extraction methods are promising techniques for the extraction of useful signals from the EEG, ECG, EMG, and PCG recorded raw data as well. The ECG/PCG data can be first decomposed into useful signal and noise subspaces using standard techniques like local and robust principal component analysis (PCA), singular value decomposition (SVD), and nonlinear adaptive filtering.

The ICA approach enables us to project each independent component (independent "cardiac acoustic source") onto an activation map at the thorax level. For each acoustic activation map, an ECG/PCG source localization procedure can be performed, looking only for a single source (or dual source) per map. By localizing multiple dipoles independently, dramatic reduction of the complexity of the computation, and increase the likelihood of efficiently converging to the correct and reliable solution.

The concept of independent component analysis (ICA) lies in the fact that the signals may be decomposed into their constituent independent components. In places where the combined source signals can be assumed independent from each other, this concept plays a crucial role in separation and denoising the signals. A measure of independence may easily be described to evaluate the independence of the decomposed components. Generally, considering the multichannel signal as $y(n)$ and the constituent signal components as $y_i(n)$, the $y_i(n)$ are independent if:

$$P_Y(y(n)) = \prod_{i=1}^{m} p_y(y_i(n)), \quad \text{for all n,} \tag{6.9}$$

where P_Y is the joint probability distribution, $p_y(y_i(n))$ are the marginal distributions, and m is the number of independent components.

An important application of ICA is in blind source separation (BSS). BSS is an approach to estimate and recover the independent source signals using only the information of their mixtures observed at the recording channels.

Due to its variety of applications, BSS has attracted much attention recently. BSS of acoustic signals is often referred to as the (cocktail party problem) [83], which means separation of individual sounds from a number of recordings in an uncontrolled environment such as a cocktail party. As expected, ICA can be useful if the original sources are independent, i.e.,

$$(p(s(n)) = \prod_{i=1}^{m} p_i(s_i(n))). \tag{6.10}$$

A perfect separation of the signals requires taking into account the structure of the mixing process. In a real-life application, however, this process is unknown, but some assumptions may be made about the source statistics. Generally, the BSS algorithms do not make realistic assumptions about the environment in order to make the problem more tractable. There are typically three assumptions about the mixing medium.

The most simple but widely used case is the instantaneous case, where the source signals arrive at the sensors at the same time.

Figure 6.6 displays the interconnection diagram of ICA-PCG signals classification escorted with localization of sound sources based on active switching of multichannel acquisition system. In addition, the auto-generated annotation mechanism will be implemented in this ICA-decomposition model too.

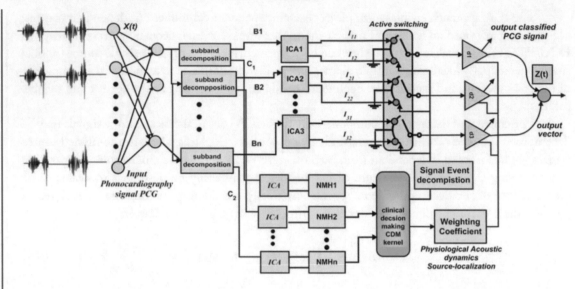

Figure 6.6: Block diagram of ICA-PCG signal classification and source localization, which use blind source separation (BSS). The internal stage between input PCG vectors X(t) and Z(t) consists of sub-band decomposition, normalization-mean data homogeneity (HNM), signal event decomposition, weighting coefficients setting, and programmable gain amplifier system at the output stage of ICA-BSS module.

This configuration was also integrated in clinical decision making profile, which will be used widely in modern medical instrumentation platforms. This has been considered for separation of biological acoustic signals such as the PCG, where the signals have narrow bandwidths and the sampling frequency is normally low. The BSS model in this case can be easily formulated as follows:

$$x(n) = H.s(n) + v(n) \tag{6.11}$$

where m $\times 1$ s(n), $n_e \times 1$ x(n), and $n_e \times 1$ v(n) denote, respectively, the vectors of source signals, observed signals, and noise at discrete time (n). H is the mixing matrix of size n_{em}. The separation is performed by means of a separating m $\times$ ne matrix, W, which uses only the information about x(n) to reconstruct the original source signals (or the independent components) as follows:

$$y(n) = W.x(n). \tag{6.12}$$

In acoustic applications, however, there are usually time lags between the arrival of the signals at the sensors. The signals also may arrive through multiple paths. This type of mixing model is called a convoluting model which is demonstrated in Fig. 6.6 where an adaptive ICA algorithm and pre-weighting function $g(y_1)$ are used together to output classified PCG vector.

One example is in places where the acoustic properties of the environment vary, such as a room environment surrounded by walls, or nearby vibration sources. Based on these assumptions, the convoluting mixing model can be classified into two more types: anechoic and echoic. In both cases, the vector representations of mixing and separating processes are changed to x(n)=H(n)∗s(n)+v(n) and y(n)=W(n)∗x(n), respectively, where (∗) denotes the convolution operation. In an anechoic

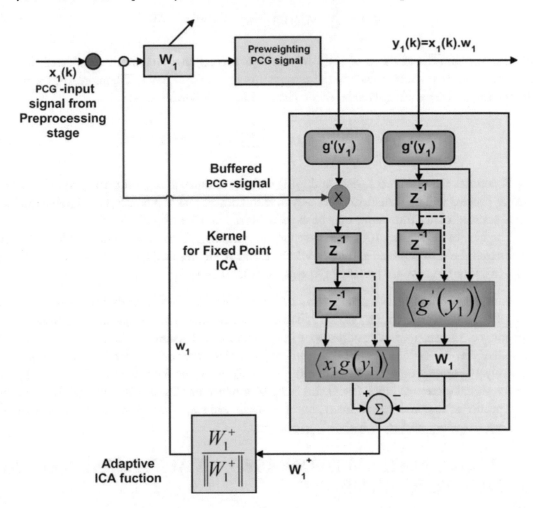

Figure 6.7: Fixed-point ICA PCG signal decomposition structure, which is used in localizing and classifying of the PCG signal vector [56].

model which is in this case the phonocardiography auscultation, however, the expansion of the mixing process may be given as:

$$x_i(n) = \sum_{j=1}^{M} h_{ij} s_j(n.\delta_{ij}) + v_i(n), \text{ for } i=1,\dots,N, \tag{6.13}$$

where the attenuation, h_{ij}, and delay, δ_{ij}, of source (j) to sensor (i) would be determined by the physical position of the source relative to the sensors. Then the unmixing process will be given as:

$$y_i(m) = \sum_{j=1}^{M} w_{ij} x_i(m - \delta_{ij}), \text{ for i=1,\ldots,M,} \qquad (6.14)$$

where the w_{ji}'s are the elements of (W). In an echoic mixing environment, it is expected that the signals from the same sources reach to the sensors through multiple paths. Therefore, the expansion of the mixing and separating models will be changed to the following mode:

$$x_i(n) = \sum_{j=1}^{M} \sum_{k=1}^{K} h_{ij}^{k} s_j(n - \delta_{ij}^{k}) + v_i(n), \text{ for i=1,\ldots,N,} \qquad (6.15)$$

where K denotes the number of paths and $v_i(n)$ is the accumulated noise at cardiac sensor (i). The unmixing process will be formulated similarly to the anechoic one. Obviously, for a known number of signal sources an accurate result may be expected if the number of paths is known.

The aim of BSS using ICA is to estimate an unmixing matrix (W) such that Y=W.X best approximates the independent sources S, where Y and X are, respectively, matrices with columns $y(n)=[y_1(n), y_2(n), \ldots, y_m(n)]^T$ and $x(n)=[x_1(n), x_2(n), \ldots, x_n e(n)]^T$.

In any case, the unmixing matrix for the instantaneous case is expected to be equal to the inverse of the mixing matrix, i.e., $W=H^{-1}$. However, in all ICA's algorithms based upon restoring independence, the separation is subjected to permutation and scaling ambiguities in the output independent components, i.e., $W=PDH^{-1}$, where P and D are the permutation and scaling matrices, respectively. ICA performance analysis results and the S_1, S_2 separation scheme is shown in Fig. 6.8 where the signal inference ration (SIR) and 3D performance matrix were presents (above) and the spectral profile of first and second heart sound of successive two heart cycle were presents (below) by defining frequency and PCG interval against time.

6.8   PCG CLASSIFICATION BASED ON ARTIFICIAL NEURAL NETWORK (ANN)

Artificial neural networks (ANN) are valuable tools used in complex pattern recognition and classification tasks. They are semi-parametric, data-driven models capable of learning complex and highly non-linear mappings. As an ANN can assimilate subtleties that may not be apparent in an explicit fashion for human analysis, there is possibility of an ANN model to improve the accuracy of diagnosis-performance in medical care problems.

More recently, investigators have attempted to perform diagnostic analysis of phonocardiograms applying the interpretive methods used by physicians. A team from the University of Paderborn (Germany) has attempted to apply wavelet analysis techniques for processing heart sounds to obtain information considered useful by physicians in auscultation [84].

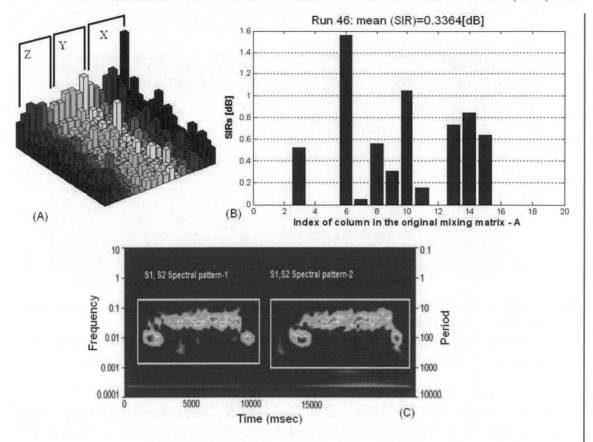

Figure 6.8: Performance results of independent component analysis (ICA-fixed point) of PCG signal, where (A) 3D performance index for signal classified according to input vectors; (B) Signal-inference-ratio (SIR) of PCG signals at -3dB attenuation diagram.

In their investigation, a neural network was utilized as an aid in processing the phonocardiogram to derive parameters of importance in clinical diagnosis. With a limited data set, the investigators were able to demonstrate the value of their approach. Other investigators have delved into joint-time frequency analysis for diagnostic assessment of phonocardiography signals [87, 88], with primary results of limited consequence. Within a neural network, the links among units are locally stored as inherent rules, either explicitly or implicitly, when they are expressed analytically. Each unit alone has certain simple properties, but when interacting with each other, such as cooperating and competing, a neural network as an entity is able to complete many complex computational tasks.

A general architecture for neural networks is shown in Fig. 6.9. The processing within a neural network may be viewed as a functional mapping from input space to output space. In principle, a unit

Table 6.1: Comparison of different phonocardiography pattern classification methods, as performance index and residual error.

Clustering Method	p-value	SIR index	PCG_mean value	cluster identified	cluster non-defined
K-Mean	0.0123	1.682	0.842	11	4
ANN-RBN	0.0167	1.732	0.732	8	4
HOS-alg.	0.0189	1.788	0.931	9	3
Basic Aggl.	0.0154	1.892	0.963	10	4
Model based	0.1923	2.013	1.038	7	4
ICA-Fixed point	0.1037	1.893	1.082	12	3
ICA-JADE	0.182	0.1028	1.820	10	2
ICA-SONS	0.161	0.1043	1.710	11	3
PCA-cluster	0.1635	1.712	1.103	8	3
Fuzzy c-mean	0.1503	1.847	1.931	7	4

in a neural network can be represented using a mathematical function, and the weights associated with the unit can be represented in forms of coefficients of that function.

The functional aggregation among different units, which creates the mapping from input to output space, is determined through both algorithm and architecture of a neural network.

6.8.1   GENERAL CONCEPT OF ANN

Since the early days of computer science it has become evident that conventional computers lack certain abilities that every human being possesses. In particular, these machines do not display a form of intelligent behavior. There have been two approaches geared at improving this situation. One is based on symbolism and the other one is based on connectionism. The former approach models intelligence in terms of computer programs which are able to manipulate symbols given a certain amount of (knowledge) and following a certain set of rules.

The connectionist approach to introducing intelligence to computer systems relies on the hope that it is possible to model the structure of the biological neural systems such as the human brain. A biological nervous system consists of a network of neurons which continually receive and transmit signals. A simple model of a biological neuron, consists of a processing element receiving several inputs.

In Fig. 6.9, the symbols $x_1, ..., x_n$ represent the strengths of (1xn) the impulses. The synaptic weights or connection strengths-denoted by the symbols $w_1, ..., w_n$ — interpret the role that the synapses play in the transmission of impulses. The output signal is represented by the symbol (y). The dependence of the output y on the inputs $x_1, ..., x_n$ is given by the following rule:

$$y = f\left(\left[\sum_{i=1}^{n} w_i . x_i\right] - \theta\right),$$

(6.16)

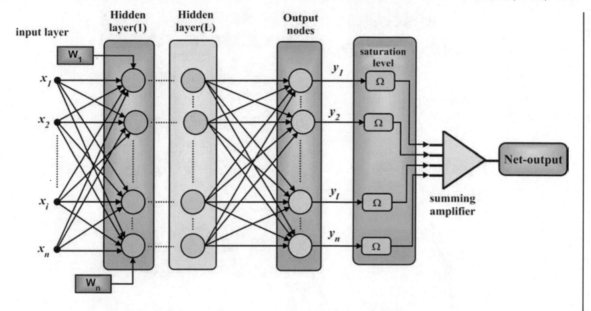

Figure 6.9: A multilayer feed forward neural network with L-hidden layer that widely used in many signal processing and pattern classification application.

where θ is a threshold value or bias and f is the neuron's activation function. One of the most commonly used activation functions is the Heaviside step function (Dirac delta function), in some terminology given by

$$f : R \rightarrow R$$

$$H(x) = \int_{-\inf}^{x} \delta(t)dt. \tag{6.17}$$

6.8.2 NEURAL NETWORK TOPOLOGIES

The neurons in an artificial neural network are sometimes also called nodes or units. The topology of a neural network refers to its framework and its interconnection scheme. In many cases, the framework of a neural network consists of several layers of nodes. The literature on neural networks distinguishes between the following types of layers:

- Input Layer: A layer of neurons which receive external input from outside the network.

- Output Layer: The layer of neurons which produces the output of the network.

- Hidden Layer: A layer composed of neurons whose interaction is restricted to other neurons in the network.

A neural network is called a single-layer neural network if it has no hidden layers of nodes, or equivalently if it has just one layer of weights. A multilayer neural network is equipped with one

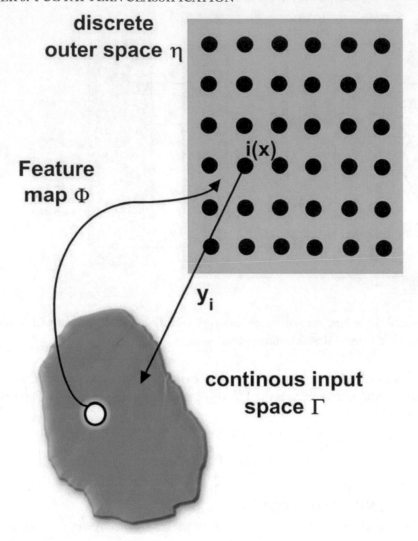

Figure 6.10: The relationship between feature map (Φ) and weight vector y_i of winning neuron i.

or more hidden layer of nodes. A feed forward neural network refers to a neural network whose connections point in the direction of the output layer.

A recurrent neural network has connections between nodes of the same layer and/or connections pointing in the direction of the input layer.

6.8.3   PCG DIAGNOSIS WITH SELF-ORGANIZING MAPPING (SOM)

In a self-organizing map (SOM), in reference to Kohonen et al. [89], the neurons are placed at the nodes of a lattice, and they become selectively tuned to various input patterns (vectors) in the course of a competitive learning process.

The process is characterized by the formation of a topographic map in which the spatial locations (i.e., coordinates) of the neurons in the lattice correspond to intrinsic features of the input patterns. Figure 6.10 illustrates the basic idea of an SOM, assuming the use of a two-dimensional lattice of neurons as the network structure. In reality, the SOM belongs to the class of vector coding

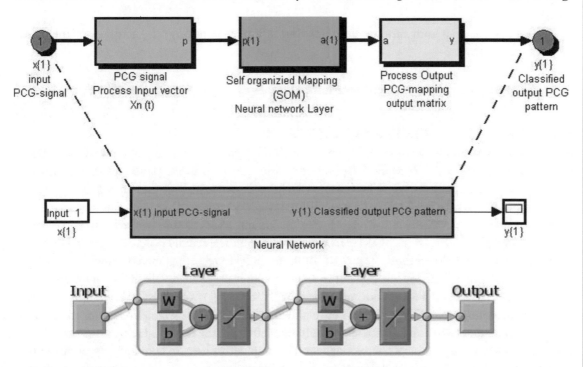

Figure 6.11: Simulink block diagram of PCG signal self-organizing mapping classification system, which uses adapted ANN-architecture for mapping different cardiac acoustic patterns. The bottom part shows the schematics for simulink realization of the SOM-network.

algorithms [90]; that is, a fixed number of code words are placed into a higher-dimensional input space, thereby facilitating data compression.

An integral feature of the SOM algorithm is the neighborhood function centered around a neuron that wins the competitive process. The neighborhood function starts by enclosing the entire lattice initially and is then allowed to shrink gradually until it encompasses the winning neuron. The algorithm exhibits two distinct phases in its operation.

1. Ordering phase, during which the topological ordering of the weight vectors takes place.

2. Convergence phase, during which the computational map is fine tuned.

Generally, the SOM algorithm exhibits the following properties.

1. Approximation of the continuous input space by the weight vectors of the discrete lattice.

2. Topological ordering exemplified by the fact that the spatial location of a neuron in the lattice corresponds to a particular feature of the input signal pattern.

3. The feature map computed by the algorithm reflects variations in the statistics of the input distribution.

4. SOM may be viewed as a nonlinear form of principal components analysis (PCA).

6.8.4   SELF-ORGANIZATION PRINCIPLE

Self-organizing in networks is one of the most fascinating topics in the neural network field. Such networks can learn to detect regularities and correlations in their input and adapt their future responses to that input accordingly. The neurons of competitive networks learn to recognize groups of similar input vectors. Self-organizing maps learn to recognize groups of similar input vectors in such a way that neurons physically near each other in the neuron layer respond to similar input vectors.

The weight matrix of the SOM can be computed iteratively as the progress of neural network training epoch were determined, Fig. 6.12 illustrates SOM weighting matrix which are used to classify PCG-signals. Self-organizing feature maps topologically emulate salient features in the input signal space at the neural network output without explicitly using supervision or even reinforcement of correct output behavior. The network's output neurons are usually conveniently arranged in single one-dimensional or two-dimensional layers. Full connectivity to the inputs is tacitly assumed. Lateral positive and negative feedback connections are also applied to help in convincingly deciding the outcome o f competitive learning. Winning a competition lets a specific output neuron reach (on state) and thus updates its weights and the weights of its surrounding neighborhood. Normalization of all weights, in addition to controlling the size of surrounding neighborhoods, usually improves the network performance by equalizing the relative changes in weight connections.

Neuron activities and interactions can be represented by a set of discrete nonlinear mathematical equations, as proposed by Kohonen [89]. Therefore, the strengths of interconnecting weights are expressed in an $n \times m$ weight matrix $W(k)$, and the lateral feed- back coefficients are similarly collected in an $n \times n$ matrix C, which has a symmetrical band structure.

Furthermore, the width of this band structure determines the effective size of neighborhoods surrounding each output neuron: Let n be the total number of output layer neurons, and let $Y(k) \in R$ be the neuron outputs at the kth discrete iteration step. Let $X(k) \in R^m$ and $U(k) \in R^n$ be the input stimuli vector and the net weighted sum. Finally, consider a nonlinear activation function designated

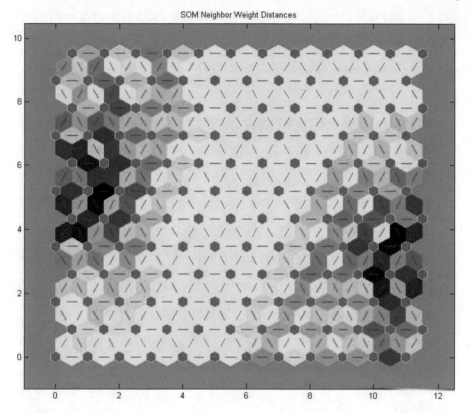

Figure 6.12: Weights matrix for self-organizing mapping (SOM) which are used to classify PCG patterns.

by: $\Phi : R^n \rightarrow R^n$. Then the output neuron activity is modeled with following:

$$Y(k+1) = \phi[V(k)] \tag{6.18}$$
$$V(k) = U(k) + \beta C(k)Y(k) \tag{6.19}$$
$$U(k) = W(k)X(k) \tag{6.20}$$

and (β) reflects a scalar relaxation factor that increases or decreases the effect of lateral feedback connections. The set of Eqs. (6.18)–(6.20), may be solved assuming typical center-surrounding input vector patterns X(k). Considerable simplification is effected if Φ is taken to be piecewise linear and if

$$C(k) = C.$$

These assumptions produce an emergent output neuron behavior that amounts to be omitted lateral feedback and using a variable-size surrounding neighborhood that depends on k.

The concept of neighborhood allows gradually decoupling topological groups of output layer neurons, which is similar to fuzzy system membership functions.

Figures 6.12 and 6.13 display the weight's plot of four PCG signal vectors by using center-surrounding input vector as reference pattern for SOM network training, which results in spatial distribution of PCCG patterns according to the intensity of the signal and its associated temporal characteristics variations. For advanced application of PCG-SOM data clustering, which can be used

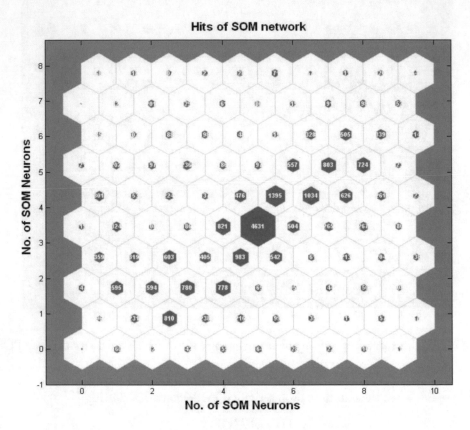

Figure 6.13: Four PCG signal pattern classification based on SOM-clustering technique, where these signal classifier using 34-layer SOM-neural network classifier system.

to identify gradations of murmurs and other heart sounds disorders, Figs. 6.14 and 6.15(a) shows the weights plot of six different PCG data vector of different valvular hemodynamic conditions. The color labeling of each PCG vector was assigned in the correlation matrix plot in Fig. 6.15(b). The weight's matrix represents the degree of probabilistic mapping of input PCG signal vectors, in which the dense areas of the plot indicate that the network energy have a high level. This also reflects the classified pattern clustered in this high energy zones.

Self-organizing maps differ from conventional competitive learning in terms of which neurons get their weights updated. Instead of updating only the winner, feature maps update the weights of

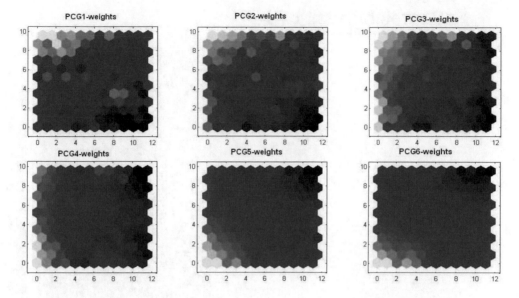

Figure 6.14: Phonocardiography SOM classification weighting mapping indicated that a shift in the linear distribution of the weights could be useful in identifying PCG dynamics variation.

the winner and its neighbors. The result is that neighboring neurons tend to have similar weight vectors and are responsive to similar input vectors. The SOM technique, appears to be a robust technique in classifying the phonocardiography signals, in spite of it's complex characteristics and computational efforts cost.

6.9   BAYES CLASSIFIER

Bayesian classifier, sometimes referred to as Naive Bayes classifier, pointed to statistics dealing with a simple probabilistic classifier based on applying Bayes' theorem with strong independence assumptions. A more descriptive term for the underlying probability model would be (independent feature model).

For simplification, a naive Bayes classifier assumes that the presence (or absence) of a particular feature of a class is unrelated to the presence (or absence) of any other feature, e.g., a ball may be considered to be an geometric-shape if it is round, and have the same diameter for each point on its surface. Even though these features depend on the existence of the other features, a naive Bayes classifier considers all of these properties to independently contribute to the probability that this shape is a ball.

Depending on the precise nature of the probability statistics model, naive Bayes classifiers can be trained very efficiently in a supervised learning scheme. In many practical applications, parameter estimation for naive Bayes models uses the method of maximum likelihood; in other words, one can

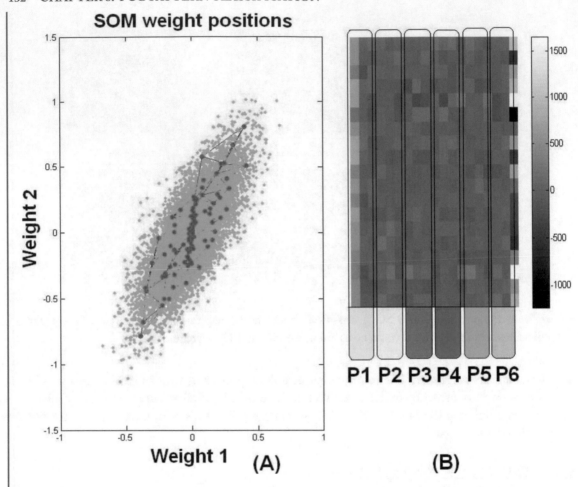

Figure 6.15: Phonocardiography weights coefficient for a training classification applied to 6-PCG vectors, where (A) is the SOM weights spatial distribution, (B)is the covariance index for the 6-PCG signals.

work with the naive Bayes model without believing in Bayesian probability or using any Bayesian methods.

6.9.1   PCG SIGNAL BAYESIAN PARAMETER ESTIMATION

All Bayes model parameters (i.e., class priors and feature probability distributions) can be approximated with relative frequencies from the training set. These are maximum likelihood estimates of the probabilities. Non-discrete features need to be discretized first. Discretization can be unsupervised (ad-hoc selection of bins) or supervised (binning guided by information in training data).

If a given class and feature value never occur together in the training set, in this case the presence of PCG S_1 and S_2 then the frequency-based probability estimate will be zero.

This is problematic since it will wipe out all information in the other probabilities when they are multiplied, when the repetition of cardiac cycle may occur in random sequence or in semi-linear profile. It is therefore often desirable to incorporate a small-sample correction in all probability estimates such that no probability is ever set to be exactly zero.

Generally speaking the derivation of the independent feature model, that is, the naive Bayes probability model.

The naive Bayes classifier combines this model with a decision rule. One common rule is to pick the hypothesis that is most probable; this is known as the maximum a posteriori or MAP decision rule. The corresponding classifier is the function classify defined as follows:

$$\Gamma(f_1, ..., f_2) = argmax_c\, p(C = c) \prod_{i=1}^{n} p(F_i = f_i \,|\, C = c) \,. \qquad (6.21)$$

In Eq. (6.21), the Γ denotes as the classification scheme of given dataset $p(F_i)$.

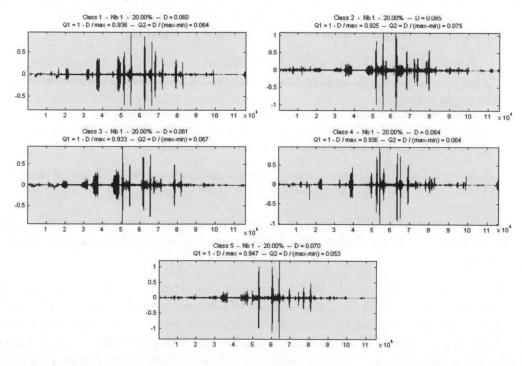

Figure 6.16: Classification scheme of phonocardiography signal illustrates five dominant heart sounds pattern.

6.10 PHONOCARDIOGRAPHY HEMODYNAMIC IDENTIFICATION

The adaptive k-mean clustering relates to a procedure for extracting information from a PCG signal obtained from a cardiac acoustic transducer and subjected to signal processing in order to aid evaluation and diagnosis of heart conditions. The method furthermore relates to techniques forming part of such extraction and apparatus to perform such feature extraction as well as coding the features to aid the ability to distinguish between related features. Signals obtained by means of a transducer are phonocardiographic representations of sounds traditionally listened to by means of a stethoscope.

Training in auscultation takes a long time and requires an aptitude for recognizing and classifying aural cues, frequently in a noisy environment. Twenty to thirty different conditions may need to be differentiated, and within each, the severity evaluated. Furthermore, there may be combinations among these. These factors contribute to explaining why not all physicians perform equally well when diagnosing heart conditions, and why it may be time-cost operation.

One of the interesting hemodynamic parameter estimation methods was achieved by phonocardiography, where the PCG signal act as a derivative for intra-cardiac pressure, by which we can considering the acceleration variable as second derivative of mechanical displacement or (pressure) and we can summarize the main aspects of this method by the following:

1. The phonocardiogram in this approach was internally recorded by using cardiovascular catheterization method, as this method will be discussed in next section in intra-cardiac phonocardiography acquisition.

2. The order of the signal derivative depends on the microphone-transducer type of mechanical energy conversion profile (displacement, velocity, acceleration).

3. There is a good similarity between the LV phonocardiogram and the LVP second derivative, and between the aortic phonocardiogram and the aortic pressure second derivative.

4. Dual integration of the PCG-signal can be useful for intra-cardiac pressure estimation, or at least may provide a good feature vector for estimating the pressure using advanced statistical technique.

The so-called first (S_1) and second (S_2) heart sound are very important markers in the assessment of a heart sound signal. These sounds are directly related to the functioning of the heart valves, in that S_1 is caused by the closure of the atrioventricular valves and contraction of the ventricles and S_2 is caused by the closure of the aortic and pulmonary valves. A number of techniques are relate to the extraction of the S_1 and S_2 signals, such as Kimberly methods and Wiener-Fassbinder technique [91], which concerns the measurement of the time interval between the S_1 and S_2 signals in relation to the heart rate in order to determine the degree of coronary artery disease. The measurement is based on peak detection and autocorrelation and it may be considered a relatively slow process.

The detection of S_1 and S_2 is obtained by performing the steps of feature extraction and classification based on the energy distribution over time in a feature time function. The feature extraction is performed by the steps of bandpass filtering, followed by instantaneous power and lowpass filtering. This generates a series of signal peaks or (hills), each relating to either an S_1 or an S_2, and a signal classification step determines which (hill) is to be regarded as either an S_1 or an S_2, whereby a systole is correctly identified a different category of signals related to various cardiac hemodynamic conditions is generally known as murmurs. The known procedures of isolating and categorizing murmurs are generally dependent on the simultaneous recording of electrocardiographic (ECG) data, by using electronic stethoscope acquisition system such as, Medtronic® and Philips® MS-Trend2300, and this complicates the practical use of auscultation techniques considerably.

The above solutions are very complex and rely on techniques that are equivalent to a long averaging time. According to the research by author, a method has been derived which is more precise and obtains a faster result. This is obtained by a sequence of steps, comprising an optional adaptive noise reduction, detection of S_1 and S_2, e.g., by means of the feature extraction procedure mentioned above, enhancement of the signal by elimination of the S_1 and S_2 contributions, performing spectral analysis and feature enhancement in order to obtain the energy content present in areas of a time-frequency representation delimited by frequency band times time interval in the form of energy distributions, classifying the energy distributions according to pre-defined criteria, and comparing the energy distributions to a paradigm of distributions related to known medical conditions and extracting information by comparing the enhanced signal to stored time functions.

The correct placement in time of S_1 and S_2 permits the energy relating to these sounds to be eliminated in the signal processing, and the resulting sound (including murmurs, regurgitation, etc.) is a useful starting signal for further analysis, because it increases the dynamic range of the remaining signal.

It also permits presenting the remaining signal to the ears with a superposition of correctly placed but (neutral) S_1 and S_2 contributions as mere time markers, but without any signal that the listener needs to process in the listening process. Diagnostic classification and evaluation is obtained by identifying specific features in order to extract characteristic patterns which are compared to a library of patterns typical of various kinds of heart disorder, and the closeness of the measured signal to these patterns.

Enhanced appreciation and identification of the heart sound features is obtained by placing the extracted features in a synthetic acoustic environment relying on supplying different signals to the ears of a listener by means of headphones. This is obtained by means of so-called Head Related Transfer Functions, or HRTF. A specific procedure of the technique is characteristic in that first and second heart sounds are detected and placed correctly on a time axis by performing the steps of feature extraction and classification based on the energy distribution over time in a feature time function by the steps of bandpass filtering, followed by instantaneous power and low-pass filtering of the original phonocardiographic signal.

The 3D-PCG feature mapping based on K-mean clustering was represents in Fig. 6.17 with the allocation of intensity variation in time and frequency domain, which have a lasting influence on wide cardiomyopathic cases, e.g., dilated cardiomyopathy (DCM), ventricular septal defect (VSD), and restrictive cardiomyopathy (RCM), by which a deviation in temporal and spatial characteristics can be observed.

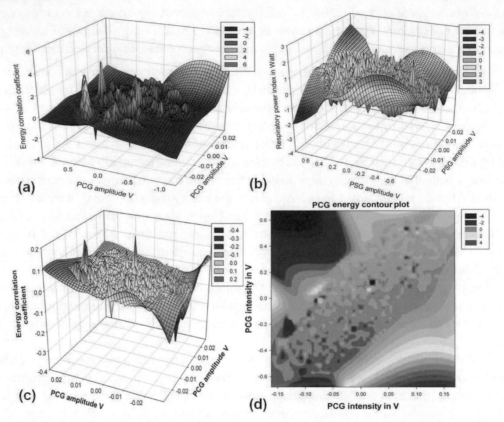

Figure 6.17: 3D representation of classified phonocardiography signals used K-mean neighborhood method: (a) normal heart sound; (b) aortic stenosis; (c) mitral valve regurgitation; (d) contouring plot of phonocardiography.

An embodiment of the techniques is particular in that it comprises the steps of extracting the first and second heart sounds by classification according to energy levels, eliminating the contribution of the said first and second heart sounds from the signal, performing spectral analysis and feature enhancement in order to obtain the energy content present in areas of a time-frequency representation delimited by frequency band times time interval in the form of energy distributions, classifying the energy distributions according to pre-defined criteria comparing the energy distributions to a descriptive distributions related to known medical conditions.

6.11 PCG PATTERN CLASSIFICATION APPLICATION EXAMPLE

The input for the procedure consists of 45 seconds of heart sound signal, sampled at a rate of 2000 Hz and read into a digital register subsequent to analog-to-digital conversion ADC. The procedure is described with reference to modern digital technology, however in principle, the various classification and sorting of time intervals and levels may be performed by analog means on DC voltages and traditional gates.

The transduction for S_1 and S_2 essentially consists of two separate processes: a feature extraction part and a classification part. The purpose of the feature extraction is to transform the input signal to a domain, in which the respective location in time of S_1 and S_2 is more distinct than in the original signal. The classification part determines the precise location of S_1 and S_2 and correctly identifies them as such hemodynamic index.

Figure 6.17(d) demonstrates how murmurs may be observed in a contouring spectrogram plotting of a time function of an original heart sounds as a correlation between PCG intensities. The spectrogram is obtained by Fast Fourier Transform. The first and second heart sounds, S_1 and S_2, have only a low-frequency content compared to the broad-band nature of the murmurs, and for this reason the signal is band-pass filtered by convolution of the original signal with the impulse response function of a bandpass filter. The corresponding spectrogram peaks of higher energy are visible but not clearly identifiable.

In order to obtain a time function of the occurrence of these higher energies, the time marginal distribution of the spectrogram is performed according to Eq. (6.13). Hereby a final PCG feature is obtained as a discrete time function. In essence, this time function, is obtained by bandpass filtering, instantaneous power extraction and lowpass filtering.

It is now clear that the final feature displays a hill every time an S_1 or S_2 occurs in the heart signal. As the magnitudes of the hills corresponding to S_1 and S_2 are comparable, it is necessary to distinguish between them by applying classification rules. First, all hills in the final feature must be identified. This is obtained for all samples of the time function which fulfill the following criteria: feature (k-l)< feature (k) and feature (k) and g_t; feature (k+1).

The next step is to construct (a) table of possible systoles. A systole is a-pair of hills (S_l and S_2) constrained by the time distance between the hills.

The time distance must fall within the following limits: 230 ms $<$ T $<$ 500 ms for human hearts.

The final sequences of systoles is determined by finding the sequence of systoles in the table having maximum energy that fulfill the following constraints:

- Systole time deviation <18%—time between systoles (diastole)> 0.9 times systole time— amplitude deviation of S_2; and

- it is 500%—amplitude deviation of S_2< 500%—in the case of overlapping systoles, the systole with maximum energy must be selected.

The result of the heart sounds identification is displayed in Fig. 6.16, in which the dominant PCG features represent in a black line with different intensity levels. These peaks of intensity annotate the time position of a first heart sound S_1 and a second heart sound S_2.

With the time positions of the first (S_1) and second (S_2) heart sounds which are correctly detected in the signal domain (given as sample numbers and a class number corresponding to positions measured in milliseconds) it is now possible to evaluate the much weaker sounds, the heart murmurs PCG traces. In the following, these detected PCG-signal time positions will be referred to as S_1 markers and S_2 markers, respectively. Reference is again made to Fig. 6.16.

6.11.1 DELIMITATION OF SYSTOLES AND DIASTOLES ASSESSMENT

Only the systole and diastole parts of the heart sound signal are used for the murmur detection. All periods, beginning 50 ms after an S_1 marker and ending 50 ms before the immediately following S_2 marker, are defined as systoles. Correspondingly, all periods, beginning 50 ms after an marker and ending 50 ms before the immediately following S_2 marker, are defined as diastoles. This is a primitive, but efficient manner of eliminating the influence of the very energetic first and second heart sounds. At later stage-in the performance of the procedure some corrections are made (side below), but it may be more advantageous to perform the elimination using more refined approaches at this early stage in the procedure.

6.11.2 TIME AND FREQUENCY DECOMPOSITION OF SYSTOLES AND DIASTOLES

The sound energy content in the sound signal is calculated by means of a spectrogram based on the Discrete Fourier Transform (DFT) using a vector length which is a power of 2, such as 16. In order to be able to classify murmurs regarding frequency contents and time distribution, each systole and diastole is decomposed into 14 frequency bands and 10 time slices, the two lowest frequency bands being discarded.

The 14 frequency bands cover the frequency range from 62.5–500 Hz, each having a width of 31.25 Hz. Before computation of the spectrogram, the sound signal is differentiated twice (corresponding to a second order high-pass filtration) in order to take into account the frequency characteristics of the human hearing, being more sensitive to higher than lower frequencies within the frequency range in question. It is considered that a parallel bank of band pass filters will perform faster in the present environment. The 10 time slices for a given systole or diastole all have the same width, corresponding to (1/10) of the total length of the systole/diastole.

The combination of frequency bands and time slices creates a 14x10 matrix for each systole/diastole. For each element in these matrices, the energy content is divided by the width of the relevant time slice, thus yielding matrices containing the heart sound power (energy per time, energy flow rate) for the 140 time/frequency elements of each systole/diastole phase [92, 93].

6.11.3 EXTRACTION OF PCG POWER AND FREQUENCY FEATURE VECTORS

A systole power (SP) vector with 10 elements is constructed by summing the 14 standard power values for each of the 10 time slices. Thus, the SP vector consists of the column sums for the S matrix. A diastole power vector (DP) is constructed in the same way. A systole mean frequency (SMF) vector (also with 10 elements) is calculated by weighting the power value for each frequency band with the mean frequency of the corresponding band, summing the 14 results, and dividing the sum with the corresponding element in the SP vector. Correspondingly, a diastole mean frequency (DMF) vector is calculated according to the power-weighting algorithm.

6.11.4 CORRECTION OF FEATURE VECTORS FOR S_1/S_2 REMNANTS

The phonocardiography feature vector (PCGFV) is one of the discriminative index for the systolic-diastolic hemodynamic instabilities. The processing method for extraction the PCGFV vector can be obtained using regressive-time delay technique. The determination of PCG features and its corresponding vectors, will vary during the online-identification of S_1 and S_2 components.

The compensation of this problem will be applied using a correlogram of both signals, which is graphical representation of the autocorrelations ρ_i versus f the time lags of the PCG trace. To verify the stability of feature-detection and quantifying the remnant error (ϵ_i) value during this process, a correction function, ($\psi(i)$), applied to the output PCG-feature vector, obtains the mean corrected S_1 and S_2 profiles.

6.12 FUTURE TRENDS IN PHONOCARDIOGRAPHY PATTERN CLASSIFICATION

A further advantageous of the automated K-mean (AKM) clustering algorithm application for extracting murmur-information is particular in that it comprises the steps of obtaining a digital representation of heart sounds for a predetermined number of seconds.

In addition to that, the identifying the time of occurrence of the first and second heart sounds in each cycle, windowing the parts of heart sounds falling between the first and second heart sounds, and second and first heart sounds, respectively. The decomposition of the signals into a predetermined first number n_1 of frequency bands, each band being decomposed into a predetermined second number n_2 of time-slices, obtaining a systole (SP) and a diastole (DP) power vector consisting of the sum of n_1 powers measured in each of the n_2 time slices, for each combination of a frequency band and a time slice.

The power values from the different systoles are compared, and the median value are chosen to be the standard value for a power vector. By obtaining a systole (SMF) and a diastole (DMF) mean frequency vector using the weighting of a power value for each of n_l frequency bands with the mean frequency of the corresponding band, summing the results and dividing the sum by the corresponding element in the respective systole or diastole power vector, while using the time of occurrence of the intensity vectors of the various classes for classifying the time distribution of murmurs.

A further embodiment of the AKM is particular in that it comprises a step preceding the step of obtaining systole and diastole murmur intensity vectors SI and DI, namely refining the windowing by setting the values of SP, DP, SMF, and DMF of the first or last elements equal to the second or last-but-one values, respectively, if the values of the first or last elements of the corresponding vectors fulfill predetermined deviation criteria.

A further embodiment of AKM is particular in that still further steps are included, namely subjecting the signal to double differentiation before decomposition, obtaining a systole (SI) and diastole (DI) murmur intensity vector, respectively, by taking the logarithm of the corresponding SP and DP vectors, classifying the obtained logarithmic vectors into murmur intensity classes, and comparing the energy distributions to a catalogue of distributions related to known medical conditions.

An apparatus for performing the basic procedure of automated PCG diagnosis system is particular in that it comprises analog-to-digital converter ADC means for converting a heart sound signal into sampled data, means for extracting the first and second heart sounds by classification according to energy levels, means for eliminating the contribution of the first and second heart sounds from the signal.

A method for performing spectral analysis, performing feature enhancement, and multiplication means for obtaining the energy content present in areas of a time-frequency representation are delimited by frequency band multiplied by time interval in the form of energy distributions means for classifying the energy distributions according to pre-defined criteria.

The comparator for comparing the energy distributions to a catalog of PCG patterns was set for a relevance to known medical conditions .

The manifestation of the Automated PCG classifier system is particular in that signal processing means are used to produce a spatial sound distribution based on frequency, a low frequency band being delivered to a first earpiece of a headphone and a high frequency band being delivered to a second earpieces of the headphone, the frequency bands containing first and second heart sounds and murmur sounds respectively.

Moreover, the diligence of the apparatus is particular in that, the signal processing means, produce a temporal sound distribution, sound signals being first delivered to a first earpiece of the headphone and then being delivered to a second earpiece of the headphone. A further embodiment of the apparatus is particular in that the signal processing comprise at least one Wiener filter module or other comparable filtering system functionality [94, 97].

6.13 SUMMARY

PCG signal pattern classification can be falls into the following categories.

- Supervised PCG pattern classification.

- Unsupervised PCG pattern classification.

- Higher-order statistics pattern classification.

- Independent component analysis (ICA) and application of mixture model ICA-MM technique.

- PCA PCG pattern classification technique.

- Self organized mapping (SOM), as a supervised adaptive classification technique.

- Bayesian classification as one of higher-order statistical approach in PCG data clustering.

C H A P T E R   7

Special Application of Phonocardiography

7.1   INTRODUCTION

Phonocardiography has an additional potential application that may be a valuable tool in primary clinical diagnosis and in cardiac monitoring functions. These applications, in some selected patients, tends to be obvious to monitor. One of these applications is analysis of the frequency of the second sound as a noninvasive indicator of sub-clinical stiffening of the aortic valve. Such analysis may be useful in identifying mild aortic stenosis. Early diagnosis in such patients may be advantageous because the prophylactic use of antibiotics is sometimes required.

Previously, we mentioned that the phonocardiography is a passive and fully non-invasive acoustic recording that provides an alternative low-cost measurement method. From this point, the development and optimization of a special clinical and medical diagnosis application, established upon phonocardiography, has been raised and grown over the past decades. In the following sections, Lighting of the main applications based on phonocardiography was achieved. Discussion of the principal and substantial PCG application occurred.

7.2   FETAL PHONOCARDIOGRAPHY SIGNAL PROCESSING

Fetal phonography is the recoding of the fetal physiological sounds. Phonocardiography, specifically, involves the acquisition and recording of fetal heart sounds. In both cases, this is achieved by sensing a fetal acoustic vibrations incident on the maternal abdomen. Fetal phonocardiography (fPCG) has a history dating as far back as the mid 19th century, and is reviewed in [87].

Up until now, the fPCG clinical data obtained using phonocardiography acquisition for fetal monitoring presented poorly in comparison with those obtained using Doppler ultrasound techniques with a research conducted by [88]. Ultrasonography currently dominates fetal heart monitoring practice.

Figure 7.1 presents the fetal phonocardiography tracing in accompanying fetal electrocardiography (fECG) which can be acquired using fECG-electrodes, and is evident to the physiological status of the fetus.

The lack of success of phonographic monitoring systems is attributed to inappropriate cardiac transducer design and physical limitation for PCG signal detection; typically, the transducers used for fetal heart monitoring were variants of those used for adults. The resulting signal had a poor signal

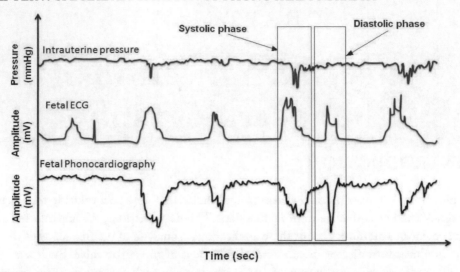

Figure 7.1: Fetal phonocardiography tracing (above) in accompaniment with fECG tracing (below) and the noisy signal can be observed in this graphical trace (Ref: Fetal clinical monitoring, 2004).

to noise ratio. This required heavy filtering which in turn led to spatial attenuation of potentially important signal information.

The design of compliance matched phonography transducers [88, 89] and represents a considerable improvement over earlier designs. Firstly, the phonography signals acquired potentially have a signal to noise ratio sufficient that heavy filtering is redundant. Secondly, these phonography sensors can characteristically operate over a wide range of fetal vibrations. Considering these factors has given rise to the concept of wide-frequency bandwidth fetal phonography.

Wide-frequency bandwidth fetal phonography enables recording of low-frequency vibrations arising from activities such as fetal breathing and movements, in addition to the higher frequency vibrations arising from heart sounds. Even though phonography microphone transducer development is at an advanced stage, adaptive and robust signal processing techniques allowing routine estimation and parameter identification of fetal activities are still blossoming. However, the advancement of the real-time signal processing would be engaged to develop linear and stable fPCG clinical diagnosis.

Figure 7.2 illustrates the data acquisition system of fPCG-signals composed of highly sensitive microphone transducer with band-pass filtering and low-pass filtering module. The purpose of this stage is to attenuate any noise contamination originated from the mother internal organs movement. Moreover, the narrow-bandwidth ADC-unit should be selected carefully to obtain accurate information from sampled fPCG-signals. Accordingly, analysis of wide bandwidth phonography and phonocardiography signals obtained using the latest transducers remains at an experimental stage.

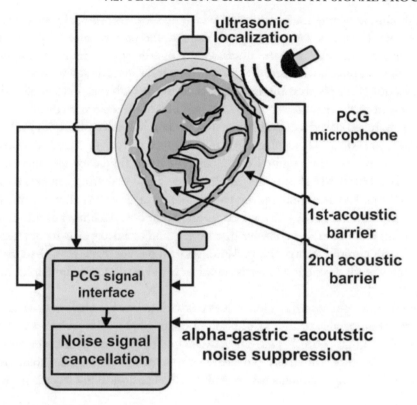

Figure 7.2: Figure schematics of fetal phonocardiography acquisition setup. The four locations of cardiac microphone to pick up phonocardiography traces of the fetus.

7.2.1   FETAL SOUND DETECTION MEDICAL SENSORS

The practice of using FHR to assess fetal status was derived from the very simple sensor in the form of the fetal stethoscope, or fetoscope. This is essentially a conical tube-geometry held in contact with the maternal abdomen, which allows a doctor or nurse to attempt to listen to sound acoustic waves produced by the fetus. Among the mixture of sounds heard, the fetal heart sounds, produced by valve opening and closing and by blood flow, may be perceived. When these heart sounds are clear, it is then possible to calculate manually the mean heart rate, say over 30-60 s.

It is also possible, with experience, to perceive changes in FHR, such as bradycardia. The use of a microphone in place of the simple fetoscope allows electronic recording of sounds detected from the maternal abdominal wall. The sounds detected include fetal heart sounds, thereby enabling the fetal phonocardiogram to be recorded [82]. However, the fetal heart sound is heard among a collection of other sounds. These may be due to movement of the fetus or the mother, maternal heart and breathing sounds, and also noises from the surrounding environment. Thus, processing of the phono-acoustic signal is essential in order to extract useful information from the fetus.

Although simple microphones were used widely in most commercially available fetal monitoring instruments, there was some attempt to improve the performance of phonocardiography by matching a suitable transducer to the mechanical properties of the tissues to which it was attached [80]. Such an approach was then used to achieve a transducer design with a wide bandwidth for fetal phonocardiography with compliance matched to the maternal abdominal wall [84]. The drawbacks arising from the limited low-frequency response of piezoelectric crystals were overcome in a compliance-matched transducer design dependent upon inductive principles [85].

The transducer INPHO has a polypropylene membrane that can be stretched to an appropriate tension in order to achieve the required compliance. It has a frequency response flat to within 3 dB from 0.1–200 Hz. It was shown that appropriate adaptive filtering can separate fetal heart sounds (>10 Hz) from fetal breathing movements (0.5–2.0 Hz) as well as eliminate the influence of maternal breathing movements [86]. It was also reported that the attachment of this transducer was best achieved using a double-sided adhesive disc rather than by means of belts or straps, since the latter interfered with compliance matching. Although fetal phonocardiography had lost popularity in recent years, new work describes a low-cost monitor based on phonocardiography and advanced signal processing [87].

The two-channel phonocardiographic device is said to provide performance for FHR variability monitoring comparable to that of ultrasound cardiography. The system was developed for home care use, offering an optimal trade-off between complexity and performance in a low-cost, stand-alone, embedded modular battery-powered instrument. The developed system provided 83% accuracy compared with the simultaneously recorded reference ultrasound records [88].

7.2.2   CHALLENGES AND MOTIVATION

Fetal phonocardiography is considered one of physiological signals which significantly can be used as clinical index for the healthy status of the fetus. Therefore, the analysis of phonocardiography faces a variety of challenges as described below.

The main challenge which faces the fetal PCG (fPCG) is the considerable acoustic impedance and scattering effect of the acoustic wave propagated from the fetus heart to the microphone transducer.

In addition to that, the principal drawback to phonocardiographically derived fetal heart recording (FHR) systems is that they are extremely sensitive to ambient noise such as maternal bowel sounds, voices in the room, certain air-conditioning systems, and especially, noise produced by any movement of the microphone or the bed clothing against the microphone.

In addition, any fetal kicking or motion produces a very loud noise that will saturate the automatic gain system on the monitor's amplifier, resulting in complete loss of recording for several seconds while waiting for the amplifier to reopen. For this reason, a manual gain control offers a great advantage when using abdominal fetal phonocardiography for recording heart rate. Furthermore, because of the high sensitivity to ambient noise, the technique is unsatisfactory for monitoring during the active phase of labor.

The current role of phonocardiographic FHR recording is quite limited but should be considered if abdominal fetal ECG and Doppler do not produce satisfactory recordings. Today, it would have to be considered below Doppler in a ranking of preferred methods of ante partum FHR recording. Both the abdominal fetal ECG and phonocardiographic FHR are rarely employed means of fetal monitoring, but are of historic significance.

The considerable features that differentiate phonography from ultrasonography and fetal electrocardiography (fECG) are the following:

1. The most up-to-date phonography transducers are able to sense fetal acoustic vibrations over a wide frequency range, and therefore can record a range of fetal activities [91].

2. Phonography is a non-invasive technique, imparts no energy to the fetus, and consequently, is inherently safe for long-term clinical monitoring of the fetus health status.

3. The modern phonography transducers are sufficiently sensitive that fetal activity can still be recorded after the position of the fetus, relative to the transducer. This is in direct contrast to Doppler ultrasound systems for which fetal movements must be tracked. In this respect, modern phonographic systems may have a reduced requirement for skilled operators.

4. The most recent phonocardiography techniques have the capability for long-term simultaneous monitoring of a range of fetal cardiac activities. This ensures that, potentially, phonography has an important role in antepartum fetal health care.

7.3   INTRACARDIAC PHONOCARDIOGRAPHY (ICP) SIGNAL PROCESSING

Intracardiac phonocardiography (ICP) is another step in the development of the scientific basis for auscultation, one of many that date from immediate auscultation and extend through echophonocardiography and nuclear cardiology diagnosis tools such as cardioSPECT and PET-imaging modalities. The 55-year history of intracardiac phonocardiography, relatively brief in duration, is quite diverse due to the interrelations with other investigative and diagnostic methods that were evolving during the same era. In the classical auscultation method, external phonocardiography, pressure manometer, and recording device development were the evolutionary forerunners of intracardiac phonocardiography, while cardiac catheterization techniques were the vehicle for implementation [89].

The ICP, as an investigative and clinically diagnostic method, has conveyed a vast amount of information which confirmed certain traditional concepts and upset others. Early enthusiastic investigators frequently overemphasized the role of ICP in diagnosis; this misplaced emphasis tended to obscure certain fundamental contributions.

In a more conservative vein, Leathaml observed that intracardiac phonocardiography has mainly served to confirm or consolidate facts which were already known, or have been ascertained in the last 50 years using multichannel external recording platform.

Yamakawa et al. used the term intracardiac phonocardiography in his study in which a condenser microphone was adapted to a vascular catheter [90], and vascular and cardiac sounds were recorded in 20 dogs and 3 humans; illustrated tracings from the animal studies were published. There were no published records of the patient studies, and apparently there were severe limitations inherent in the initial methodology.

Three recorded heart sounds and murmurs in the lesser circulation with a barium titanate transducer and noted technique was capable of localizing heart sounds and murmurs to an extent not previously possible. The frequency response of the catheter preamplifier system was linear over the range of heart sounds; however, since the response of the barium titanate dropped off sharply below 10 Hz, it was not possible to record simultaneous intracardiac pressures waveform.

A double lumen catheter was used to record pressure with an external manometer. These and additional studies in acquired and congenital heart disease were the initial ICP studies in the United States. The instrumentation and usage gave the entire field a worldwide impetus.

Figure 7.3 displays different vital parameters with mitral stenosis ICP traces where the above

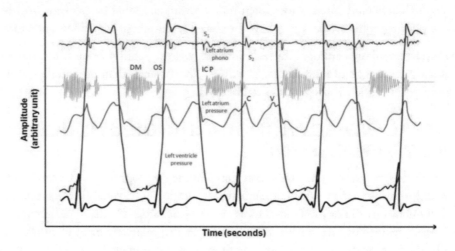

Figure 7.3: Double manometer study. Mitral stenosis (MS), sinus rhythm (SR). Top to bottom: intra cardiac phonocardiogram, left atrium, intra-cardiac phonocardiogram, left ventricle; left atrial pressure pulse; left ventricular pressure pulse; electrocardiogram, lead II. Time lines 40 ms. Paper speed 100 mm/s.(Ref:AHA 1976.)

trace is the ICP signal of the left atrium together with surface phonocardiography signal recorded from auscultation site. One can observe the capability for use in ICP tracing as invasive clinical index for cardiac pathological diagnosis procedure, but still not adequate due to several acoustic and noise contamination problem. Luisada et al. [115] used a simplified detection method of transducing pressure obtained through standard catheter systems to isolate pressure frequencies in the sound range. The technique, fluid column transmission, accelerated access to recording sound in the left heart, and was adapted or modified in a number of laboratories. An important outgrowth of these studies was the proposal by [88] for revision of the classic auscultatory areas.

The auscultatory areas were renamed according to the chamber or vessel in which a given sound or murmur was best recognized by intracardiac phonocardiography. Murmurs were further described as originating in the inflow or outflow tracts of the left and right heart. Grant's earlier physiologic concepts of ventricular inflow and outflow tracts were coupled to this anatomic and auscultatory framework.

It is possible to take the matter a step further; for example, Leatham's classification of murmurs is easily adapted to this approach. The principals gained from intra-cardiac phonocardiography provide a solid basis for the practice and teaching of cardiac auscultation technique.

7.3.1   ICP MEASUREMENT DEVICE

An inductance micro-transducer for intravascular pressure was described by Gauer and Gienappl and its evaluation and application were presented in 1951 [92]. The Allard-Laurens variation consisted of a single coil with a central core suspended between two diaphragms; displacement altered the self-inductance of the coil in a linear pattern. The experimental setup of ICP-signal clinical measurement system as a part of catheterization module was illustrated in Fig. 7.4, which describes four basic intra-cardiac inlets and cardiac microphone mounted in the wall of Swan-like catheter to detect the ICP-acoustic signals. Soulie et al. [94] presented an extensive experience with simultaneously recorded intra-cardiac sound and pressure (ICSP) in acquired and congenital heart disease. Soulie re-emphasized the contribution of turbulence in the production of certain murmurs, and anticipated the flow velocity measurements of the 1970s. He considered extra-cardiac auscultation over the thorax to be a resultant of complex and disjointed vibratory phenomena, the components of which had timing that was different from one cavity to the other. Murmurs recorded in a cardiac or vascular chamber were propagated in the direction of the jet or turbulent flow which produced them, and they tended to stay localized in the cavity of their origin.

Additionally, the use of ICP measurement profile can also be applied to identify the arterial pressure sound patterns which can be quantified through gradual increase of the arterial sound pressure (Pressure dyne PD) in variable cardiac valvular disorders ranging from aortic insufficiency to sever aortic stenosis. Figure 7.5 presents the different recorded trace of ICP signal as a function of intra-arterial sound pressure level. Millar [91] developed catheter mounted pressure transducers constructed from a pair of matched, unbounded silicon elements, geometrically arranged to act, respectively, in compression and extension upon application of pressure. The improvement of the intrinsic gauge sensitivity, thermal stability, drift characteristics, linearity, and mechanical durability represented a significant technical advance in catheter-tip pressure transduction mechanism.

7.3.2   ICP SIGNAL PROCESSING

The main problem of ICP signal is the artifact production and recognition has existed throughout the history of ICSP recording, but has diminished with each technological development. Careful

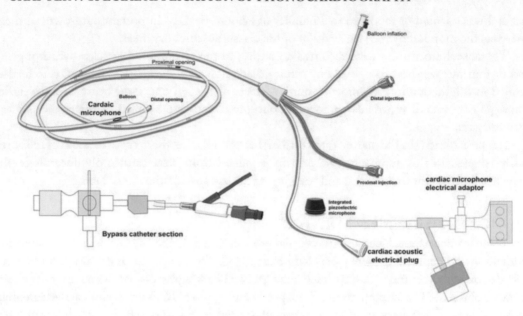

Figure 7.4: ICP-clinical instruments setup with four basic intra-cardiac inlets and the cardiac microphone unit that hinged with Swan-type catheter inserted through vascular route to record internal acoustic vibration of myocardium.

auscultation prior to study, correlation during study with external phonocardiography and echocardiography, amplification of ICP during the study, a careful search for reproducibility of recorded phenomena, and correlation of recordings with increasingly refined catheterization-angiographic and echophonocardiography techniques have reduced the potential for artifact error. The direct calibration of intracardiac cardiovascular sound was performed by Soulie et al. [94] with a cardiac acoustic transducer calibrated in units of SI-pressure (mmHg).

The signal processing structure for the ICP components was presented in Fig. 7.6, where the ICP signal is low-pass filtered, and the baseline recordings buffered actively. Succeeding that the fast Fourier transform (FFT) applied to the buffered signal. Furthermore, the valvular components can also be identified. The four specific heart sounds, S_1, S_2, S_3, and S_4, can also be extracted using feature-vector computation based on radial basis artificial neural network (RBANN)-system. Moreover, the accompanied pressure profile of the left atrium and ventricle was determined in this computational schema.

7.3.3 ICP ACOUSTIC TRANSMISSION PROPERTIES

Feruglio [93] and Silin [98] reviewed the available techniques for detecting intra-cardiac acoustics and introduced the vibrocatheter (a catheter with lateral opening near the tip covered with a thin latex cuff connected to the tubing of binaural stethoscope for direct auscultation, or to a piezoelectric

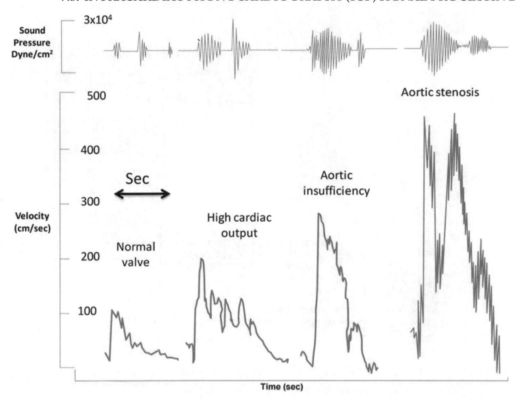

Figure 7.5: Intra-arterial sound and instantaneous aortic velocity in a patient with a normal aortic valve and cardiac output (5.4 L/min), a patient with a normal valve and high cardiac output (12.9 L/min), a patient with aortic insufficiency, and a patient with aortic stenosis. The point velocities were redrawn on an identical scale to show clearly the relation between velocity, turbulence, and intra-arterial sound. It is apparent that as blood velocity and turbulence increased, arterial sound also increased from normal 100 P.D. (Reprint from Stein, 1978, [96].)

microphone to record the intra-cardiac phonocardiogram). Observations in valvular and congenital heart disease with this system were similar to those obtained with the methods described earlier in this book.

The innovative use of the system consisted of delivery of sounds of known frequency content or intensity into the heart; the vibrocatheter was connected to a magneto-dynamics unit of an acoustically insulated loud speaker connected to a variable-frequency oscillator or tape recorder. The artificial sounds delivered into the heart were then recorded from chest surface sites in order to study the modifying effects of sound transmission from cardiac chambers to chest wall.

The attenuation of cardiovascular sound depended on the site of intra-thoracic production, the frequency components of the sounds, and the characteristics of the conducting tissues. The maximal

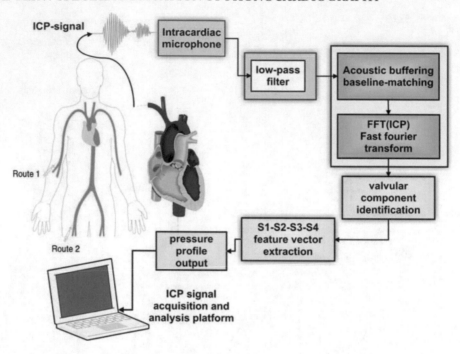

Figure 7.6: Schematic of ICP signal acquisition and processing, where the input transducer is the intra-cardiac microphone, and low-pass filtering is used to remove parasitic ambient noise (from near lung activity and large vessels blood flow. The baseline matching to avoid any amplitude drift in the ICP signals is an FFT step performed before valvular component identification. The feature-extraction kernel used to separate S_1, S_2, S_3, and S_4; the analysis results displayed on medical certified laptop platform.

attenuation occurred, when sound originated in the right and left pulmonary arteries; attenuation was less when sound arose in the right atrium and main pulmonary artery; sound was well conducted to the chest wall from the right ventricle.

The distance from sound source to chest wall and the interposition of poorly conducting tissues were considered to be the reasons for the different degrees of attenuation. Frequencies below 100 Hz and above 350 Hz were greatly attenuated; frequencies of about 200 Hz were conducted with little attenuation (presumably because these frequencies were in the same general range as the natural frequency of the thorax).

7.4 SEPARATION OF PHONOCARDIOGRAPHY FROM PHONOSPIROGRAPHY SIGNAL

Lung sounds produce an incessant noise during phonocardiography recordings that causes an intrusive, quasi-periodic acoustic interference that influences the clinical phonocardiography auscultatory interpretation. The introduction of pseudo-periodicities, due to lung sounds overlapping, mask the relevant signal and modifying the energy distribution in the PCG spectral band, in order to conduct a proper PCG-signal analysis [97]. This lung sound interference problem needs an effective reduction of this sound parasitic effect-over the disturbed phonocardiography signal to yield a successful heart PCG interpretation and identification. Generally, there are three approaches for lung sound cancellation which are as follows:

- Low-pass filtering (LPF) of the PCG signal, which attenuate the low-frequency component of lung sound.

- Application of autocorrelation function (ACF) to cancel down the redundancy harmonic of the phonospirography signals.

- Online subtraction of two successive S_1 and S_2 PCG signal with time delay (T_h) from each other and multiply with constant gain level.

The simplest technique, which is illustrate in Fig. 7.7, can be implemented in a hardware-approach based on adaptive IIR-filter that added the estimated noise-source with PCG-source to cancel out this noise effect. This versatile approach is flexible in reverse direction, where the required task is to segregate lung sound (LS) as dominant signal from PCG-signal. The other technique which is used

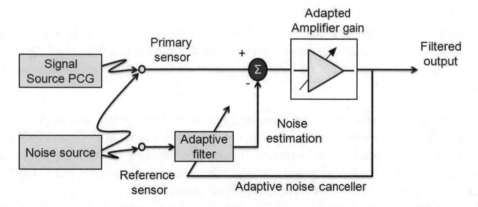

Figure 7.7: Adaptive noise cancellation method, for separation and suppression of lung sound from heart sound. The cancellation method depends on the adaptive IIR-filtering and parametric estimation algorithm to remove progressive overlapping source (e.g., lung sound) to take part in signal acquisition loop.

for suppression respiratory acoustic vibration from phonocardiography signals is by using adaptive cancellation. Adaptive noise canceling relies on the use of noise canceling by subtracting noise from a received signal, an operation controlled in an adaptive manner for the purpose of improved signal-to-noise ratio (SNR). Ordinarily, it is inadvisable to subtract noise from a received signal, because such an operation could produce disastrous results by causing an increase in the average power of the output noise. However, when proper provisions are made and filtering and subtraction are controlled by an adaptive process, it is possible to achieve a superior system performance compared to direct filtering of the received signal [95].

Basically, an adaptive noise canceler is a dual-input, closed-loop adaptive feedback system as illustrated in Fig. 7.8. The two inputs of the system are derived from a pair of cardiac microphone

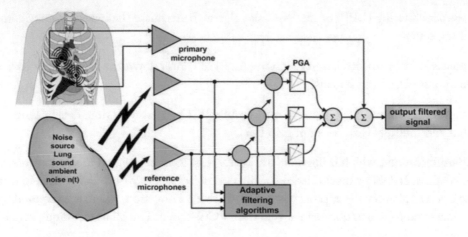

Figure 7.8: Block diagram of adaptive filtering that is used to eliminate the parasitic noise and disturbances which affect and infer the PCG signal; the noise source here assigned to the lung sounds (LS) and other ambient noise sources originated nearby the cardiac microphone. The PCG signal is to be amplified in programmable gain amplifier (PGA) in order to equilibrate the intensity drift during PCG acquisition.

sensors: a primary sensor and a reference (auxiliary) sensor. Another variation of respiratory sound cancellation is based on recording the two-acoustic traces via modeling PSG and PCG signals coincidentally to separate each other. Moreover, the clinical pulmonary function can also be derived from the acquired lung sounds, and in addition, the other buffering parameters can be used for the calibration purposes of the PSG-signal and PCG signal together.

This concept is presented in Fig. 7.9. The use of respiratory acoustic (lung sound) signal to calibrate against phonocardiography signal has been used to cancel the effect of lung sound, which propagate with PCG signal. Specifically, the following assumptions were derived:

1. The primary sensor receives an information-bearing signal s(n) corrupted by additive noise $v_0(n)$, as shown by the following formula:

$$d(n) = s(n) + v_0(n). \tag{7.1}$$

The signal s(n) and the noise $v_0(n)$ are uncorrelated with each other; that is,

$$E[s(n)v_1(n - k)] = 0 \quad \text{for all k,} \tag{7.2}$$

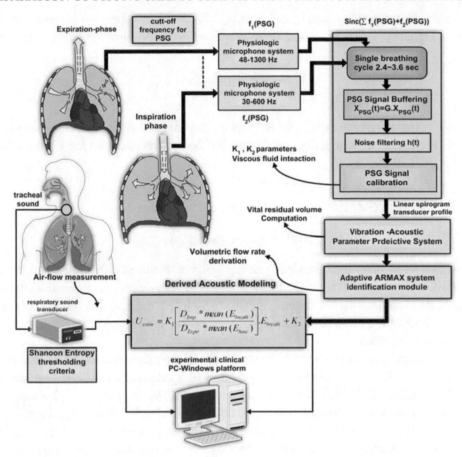

Figure 7.9: The schematic of cardio-respiratory sound recording and modeling loop. This system is used to conduct a correlation analysis and system identification between cardiac acoustics and respiratory acoustics, in which the tracheal sound and lung sounds recording synchronously to be used in calibration process for phonocardiography [101].

where s(n) and $v_0(n)$ are assumed to be real valued.

2. The reference sensor receives a noise $v_1(n)$ that is uncorrelated with the signal s(n), but correlated with the noise $v_0(n)$ in the primary sensor output in an unknown way; that is,

$$E[s(n)v_1(n-k)] = 0 \quad \text{for all k} \tag{7.3}$$

and

$$E[v_0(n)v_1(n-k)] = p(k), \tag{7.4}$$

where, as before, the signals are real valued, and p(k) is an unknown cross-correlation for lag (k). The reference signal $v_1(n-k)$ is processed by an adaptive filter to produce the output signal

$$y(n) = \sum_{k=0}^{M-1} \hat{W}_k(n)v_1(n-k), \tag{7.5}$$

where $\hat{W_k}(n)$ is the adjustable (real) tap weights of the adaptive filter. The filter output y(n) is subtracted from the primary signal d(n), serving as the "desired" response for the adaptive filter. The error signal is defined by:

$$e(n) = d(n) - y(n). \tag{7.6}$$

Thus, substituting Equ. (7.6) into Equ. (7.7), we get the following:

$$e(n) = s(n) + v_0 - y(n). \tag{7.7}$$

The error signal is, in turn, used to adjust the tap weights of the adaptive filter, and the control loop around the operations of filtering and subtraction is thereby closed. Note that the information-bearing signal s(n) is indeed part of the error signal e(n). The error signal e(n) constitutes the overall system output. From Equ. (7.7), the observer can see that the noise component (lung sound) in the system output is $v_0(n)-y(n)$.

Now the adaptive filter attempts to minimize the mean-square value (i.e., average power) of the error signal e(n). The information-bearing signal s(n) is essentially unaffected by the adaptive noise canceler. Hence, minimizing the mean-square value of the error signal e(n) is equivalent to minimizing the mean-square value of the output noise $v_0(n)-y(n)$. With the signal s(n) remaining essentially constant, it follows that the minimization of the mean-square value of the error signal is indeed the same as the maximization of the output signal-to-noise ratio of the system.

The other method of PSG and PCG-signal separation is based on referential respiratory sound recording, based on synchronized expiration-inspiration phase cancellation. This technique, used 4 auscultation site with one tracheal sound track record and mapping the two-phase component into two time-varying functions with defined bandwidth (BW) $f1_{PSG}$ $f2_{PSG}$ which are temporally summed with sinc-based window.

Through the next step, to the noise filtering and PSG calibration for parameter identification of lung viscous properties, the resultant predictive parameters were fed to the ARMAX system identification module. The derived acoustic properties, together with air flow signal, were traced from the tracheal segment used in active separation of the heart sound from lung sound. The effective use of respiratory sound adaptive noise canceling, therefore, requires that positioning the cardiac microphone reference sensor in the noise field of the primary sensor with two specific objectives in mind. First, the information-bearing signal component of the primary sensor output is undetectable in the reference sensor output. Second, the PCG reference sensor output is highly correlated with the noise component of the primary sensor output. Moreover, the adaptation of the adjustable filter coefficients must be near the optimum level.

7.5   PHONOCARDIOGRAM CARDIAC PACEMAKER DRIVEN SYSTEM

In a normal heart, the sinus node, the heart's natural pacemaker, generates electrical signals, called action potentials, which propagate through an electrical conduction system to various regions of the heart to excite myocardial tissues in these regions. Coordinated delays in the propagations of the action potentials in a normal electrical conduction system cause the various regions of the heart to contract in synchrony such that the pumping functions are performed efficiently. Thus, the normal pumping functions of the heart, indicated by hemodynamic performance, require a normal electrical system to generate the action potentials and deliver them to designated portions of the myocardium with proper timing, a normal myocardium capable of contracting with sufficient strength, and a normal electromechanical association such that all regions of the heart are excitable by the action potentials.

The function of the electrical system is indicated by electrocardiography (ECG) with at least two electrodes placed in or about the heart to sense the action potentials. When the heart functions irregularly or abnormally, one or more ECG signals indicate that contractions at various cardiac regions are chaotic and unsynchronized [99].

Such conditions, which are related to irregular or other abnormal cardiac rhythms, are known as cardiac arrhythmias. Cardiac arrhythmias result in a reduced pumping efficiency of the heart, and hence, diminished blood circulation.

Examples of such arrhythmias include bradyarrhythmias, that is, hearts that beat too slowly or irregularly, and tachyarrhythmias, which is disturbance of the heart's rhythm characterized by rapid and irregular beating. A patient may also suffer from weakened contraction strength related to deterioration of the myocardium. This further reduces the pumping efficiency.

For example, a heart failure patient suffers from an abnormal electrical conduction system with excessive conduction delays and deteriorated heart muscles that result in asynchronous and weak heart contractions, and hence, reduced pumping efficiency, or insufficient hemodynamic performance.

The phonocardiography recording during such a case is displayed in Fig. 7.10 A cardiac rhythm management system includes a cardiac rhythm management device used to restore the heart's pumping function, or hemodynamic performance. Cardiac rhythm management devices include, among other things, pacemakers, also referred to as pacers. Pacemakers are often used to treat patients with bradyarrhythmias. Such cardiac pacemakers may coordinate atrial and ventricular contractions to improve the heart's pumping efficiency. Cardiac rhythm management devices may include defibrillators that deliver higher-energy electrical stimuli to the heart.

Such defibrillators may also include cardioverters, which synchronize the delivery of such stimuli to portions of sensed intrinsic heart activity signals. Defibrillators are often used to treat patients with tachyarrhythmias. In addition to pacemakers and defibrillators, cardiac rhythm management devices also include, among other things, devices that combine the functions of pacemakers

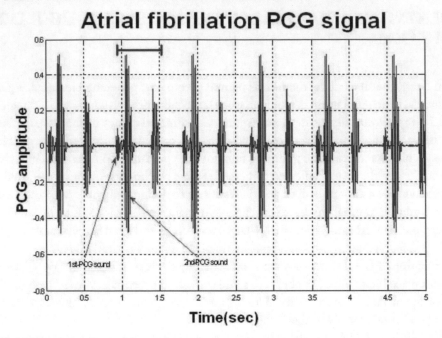

Figure 7.10: PCG traces during atrial fibrillation state, which can be used to extract timing scheme for driving pacemaker stimulation rate. This technique may have a potential application in cardiac embedded pacing and monitoring.

and defibrillator, drug delivery devices, and any other devices for diagnosing or treating cardiac arrhythmias.

The effectiveness of a cardiac rhythm management therapy is measured by its ability to restore the heart's pumping efficiency, or the hemodynamic performance, which depends on the conditions of the heart's electrical system, the myocardium, and the electromechanical association. Therefore, in addition to the ECG indicative of activities of the heart's electrical system, there is a need to measure the heart's mechanical activities indicative of the hemodynamic performance in response to the therapy, especially when the patient suffers from a deteriorated myocardium and/or poor electromechanical association.

For these and other reasons, there is a need for evaluating therapies by monitoring both electrical and mechanical activities of the heart, to give a comprehensive prospect for the actual status of the heart during pacing therapy [94, 98, 99].

7.6   BASIS OF CARDIAC SUPPORTIVE DEVICE

Most cardiac medical devices that automatically detect events in the cardiac cycle, utilize the ECG signal as the main source of information about the heart activity. Although the ECG signal represents the operation of the electrical conduction system that triggers the contraction of the heart muscle, it has several limitations.

The heart sounds or the recorded phonocardiography (PCG) traces, being a direct expression of the mechanical activity of the cardiovascular system, is potentially an additional source of information for identifying significant events in the cardiac cycle and detecting non-regular heart activity.

The spectrum application of the PCG signal in the cardiac pacing outlook is illustrates in Fig. 7.11, where the utilization of acoustic sensing component in the external cage of the pacemaker-

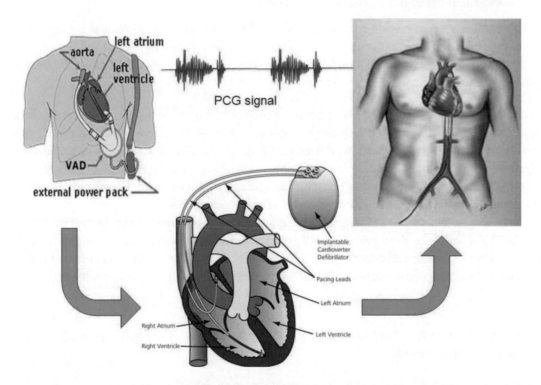

Figure 7.11: Schematic of the PCG-signal processing application in pacemaker driving system, where PCG signal can play a role as a trigger event for cardiac assist device (CAD), implantable cardiac pacemaker, and as navigator-guidance for intra-vascular catheterization.

system is used as a hemodynamic derived signal to actuate pacemaker responsively. The principal application of PCG signal in automated cardiac assisted device can be listed as below:

- A left ventricular assist device (LVAD) is a battery-operated, mechanical pump-type device that's surgically implanted. It helps maintain the pumping ability of a heart that can't effectively work on its own.

- An Intra-Aortic Balloon Pump (IABP) is a small device that is placed in the thoracic aorta, and uses both ECG and aortic pressure waveforms to time inflation and deflation of a balloon that increase or decrease the aortic pressure, and thus improves the blood flow to the arteries and reduces the workload of the heart

- Implantable cardioverter defibrillator (ICD) and automatic external defibrillator (AED) are devices that sense the cardiac rhythm, monitor, and treat life-threatening arrhythmia such as ventricular tachycardia or fibrillation. When such abnormal rhythm is detected, the device shocks the heart to restore the normal rhythm.

  The productive research direction toward the development of the phonocardiography based on adaptive ICD will turn the concept of invasive sensing mode to the non-invasive mode by the use of wireless data communication between the microphone-terminal unit and pacemaker base receiving unit.

A cardiac rhythm management system provides a phonocardiographic image indicative of a heart's mechanical events related to hemodynamic performance to allow, among other things, diagnosis of cardiac conditions and evaluation of therapies treating the cardiac conditions. The phonocardiographic image includes a stack of acoustic sensor signal segments representing multiple cardiac cycles. Each acoustic sensor signal segment includes indications of heart sounds related to the heart's mechanical events and representations of the heart's electrical events.

The diagnosis and/or therapy evaluation are performed by observing or detecting at least an occurrence of a particular heart sound related to a cardiac time interval or a trend of a particular time interval between an electrical event and a mechanical event over the cardiac time interval [97, 99]. The main architecture of the phonocardiographic pacemaker driven system consists of the following:

(A)-Pacing pulses delivery system to the heart muscle, which comprises:

- electrical sensing circuit to sense a cardiac signal; a pacing-therapy(stimulation) circuit to deliver the pacing pulses;

- vibro-acoustic sensor to produce an acoustic sensor signal indicative of heart sounds;

- supervisory controller coupled to the therapy circuit, which includes:

  A- Therapy protocol synthesizer adapted to generate a sequence of pacing parameters.

  B- An automatic therapy protocol execution module adapted to time pacing pulse deliveries each associated with one parameter of the sequence of pacing parameters.

  C- Processor coupled to the sensing circuit, the acoustic sensor, and the controller, the processor including an image formation module adapted to produce a phonocardiographic image based on the cardiac signal and the acoustic sensor signal.

The accoustince-derived signal from the microphone sensor mounted on the pacemaker cage will be further processed and analyzed using FFT-processing module. Moreover, the selective band-with IIR filter can be used to compensate possible drift in the received acoustic signal during cardiac cycle phases.

(B)-The phonocardiographic derived-image including a stack of segments of the acoustic sensor signal aligned by a selected type of the cardiac events and grouped by at least parameters of the sequence of pacing parameters.

7.7   SUMMARY

To conclude this chapter, we summarize the following points.

1. The phonocardiography can be used as a mutual tool in many clinical and biomedical engineering application such as blood pressure measurement technique, cardiac pacemaker, intra-cardiac vital parameters, and hemodynamic diagnosis instruments.

2. The ability to use fetal phonocardiography (fPCG) in the diagnostic index of the fetus has been discussed and evaluated through many trials and the synchronization of PCG records with other vital signals was also set.

3. Intra-cardiac phonocardiography signal recording and analysis is a challenging technique that has many complications and obstacles to replace other invasive-parameters. The ICP signal is highly sensitive to noise.

4. Separation of heart sound (HS) from lung sound (LS) also composed of determination of two different signals that may interfere with the detection process for both of them. Adaptive noise cancellation is considered as a steady method to solve such problem.

5. Promotive researches toward developing a new sensing mechanism for implantable cardiac pacemaker as the basis for adaptive rate responsive pacemaker system was discussed. The evolution of such sensing technique based on cardiac acoustic waveform, i.e., PCG signal will make the stimulus response of the pacemaker to behave in linear patterns This method is still under development and it is nascent technology in the cardiac pacing therapy. Although, it will assist in formulating a potential pacing-sensing approach.

CHAPTER 8

Phonocardiography Acoustic Imaging and Mapping

8.1 INTRODUCTION

The nature of phonocardiography is an acoustic vibratory signal; therefore, one of the perspective application of this signal is to develop a novel technique to synthesis a multichannel-acoustic imaging platform based on real-time synthesized acquisition (RTSA), in order to reconstruct a medical readable anatomical and physiological images.

This technique, also denoted as acoustic camera module, in which its function is to localize cardiac acoustic vibration sources. The functionality of this acoustic image was to identify the hemo dynamic turbulences and abnormal blood flow patterns associated with varieties of cardiovascular pathologies. In this chapter, the possible road map to build up and express the principals for such a type of medical imaging technique will be discussed and illustrated.

The previous chapters have intensively illustrated the benefits and horizons of the phono-cardiography application, which assists the physician and cardiologist in dealing with varieties of cardiac sounds patterns. It is based on the time domain visualization of cardiac acoustic signals.

The broad idea of this approach has been adopted from the ECG temporal and spatial analysis, where the time domain electro-physiological waveform is analyzed and reconstructed by specialized electrical impedance tomography methods such as wavelets transformation, adaptive beam-forming algorithms, and other related acoustic image formation and reconstruction methods. However, such an approach is not appropriate for acoustic signals because listening to the signal (i.e., auscultation) differs from viewing the time domain waveform, especially since acoustic events may happen simultaneously at different frequency ranges.

Cardiologists have tried to see in the waveform what should be heard (since the extensive cardiac auscultation knowledge, gathered over nearly 200 years, describes the cardiac acoustical phenomena). For this reason, the phonocardiography technique was, and still is, rejected by the medical community, although some of the trails have proved that it can be a competitive diagnostic technique from the cost point of view and the methods of analysis.

Phonocardiography acoustic tomography imaging (PATI) may be improved by multi-band analysis, multi-sensor array signal processing where several waveforms related to specific sub-bands are filtered out and often processed in a non-linear mode. Such improvements allow physicians and biomedical engineers to identify and reconstruct acoustic events related to different frequencies.

Conversely, this approach is still inadequate due to nonlinear complexity of sound perception and detection, and indeed due to the lack of adequate acoustic tomography image reconstruction algorithms.

Cardiac energy propagation in human body tissue in the audible frequency, range from 101–103 Hz which is equivalent to 900 Hz of bandwidth, which has a considerable shear-wave component when contrasted to compression waves in the ultrasound band from 104–107 Hz. This dispersive, shear wave propagation is characterized by wavelengths on the order of a few centimeters. Therefore, a multiple wavelength array aperture and frequency partition signal processing make it reasonable to compensate for frequency-dependent wave speed and form images with an aperture near-field focused beam that scans the chest volume.

The use of passive listening (auscultation) techniques to recognize arterial and cardiac valve unstable blood flow has been previously suggested for the detection of both coronary and cranial blood flow pathologies and circulation lesions.

In recent times, the opportunity of locating such arterial irregularities using near-field focused beam forming techniques has been suggested as a procedure that would enhance cardiac auscultation performance and provide a non-invasive, diagnostic screening device [102]. For coronary arterial diseases and heart valves problem, this technique requires a positioning of an array of spatially diverse cardiac microphone sensors on the external chest wall near the region of interest and situated with moderately unobstructed paths through the intercostals spaces between the ribs of the subject to the location of the turbulence. The sensor outputs are processed to generate an image of the vibration energy field in the volume beneath the acoustic sensor array.

The process locates the vibration source by back propagating the sensor signals using beam formation methods. The vibration source mapping itself would be further processed using adaptive radon transformation. This inverse acoustic mapping of the sensor outputs requires an assumed average shear-energy wave speed [100, 101] and the wave equation-based model for the propagation including geometrical Euclidean distance transit time, a homogeneous medium, and a geometric distribution (loss) model.

Figure 8.1 presents the block diagram of the cardiac acoustic imaging system based on real-time phonocardiography signal acquisition, with a defined number of cardiac microphones placed on the circumstance of the wearable textile embedded electrodes.

8.2   MOTIVATION AND PROBLEM FORMULATION

The main problems that face the researcher in the field of acoustic imaging is the transmission acoustic impedance and the scattering effect of the acoustic waves from their sources. To avoid such a complication the first hint is to use high-sensitivity acoustic transducer or sensor to increase the SNR ratio in the post-processing stage. The first trail on cardiac acoustic mapping was done by Kompis et al. [102] in which they developed an acoustic array platform with simultaneous recording of 8-16 microphone elements.

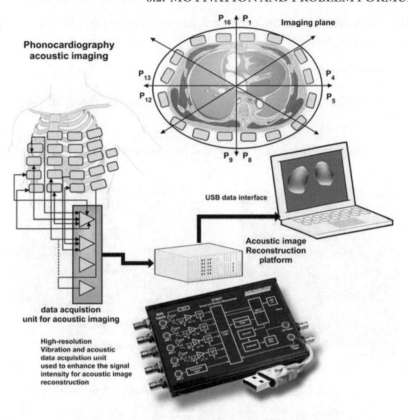

Figure 8.1: Block diagram of technical experimental setup for phonocardiography acoustics tomography imaging (PATI), which shows the configuration of cardiac microphones array on the subject chest with high precision acoustics and sound data acquisition platform (e.g., as the Data Translation DT-9837 DAQ-system for sound and vibration measurement), to enhance and accelerate acoustic image reconstruction rate.

The concurrent data acquisition mode of PATI-technique faces many problems, in the sense of sampling rate, filtering bandwidth, the synchronization of microphone signal ADC-conversion and reconstruction time. These impediments should be considered in future optimization and enhancement of the PATI-method. The anatomical-thorax model is composed of five different components of substantially different acoustic characteristics; this is effectively be illustrated in Fig. 8.2, where these components can be distinguished as follows:

- Myocardial tissue, which consists mainly of the heart muscle and associated coronary vasculature networks and the aortic components.

- Respiratory-airways which consist of bronchial tree, inferior part of larynx, and pleural space, including pleural fluid.

- Lung lobules with their parenchyma-compartment.

- The rib-cage which represents the grounded-compartment of thorax and the cardiac acoustic model.

- Muscular compartment (pectoralis major and pectoralis minor).

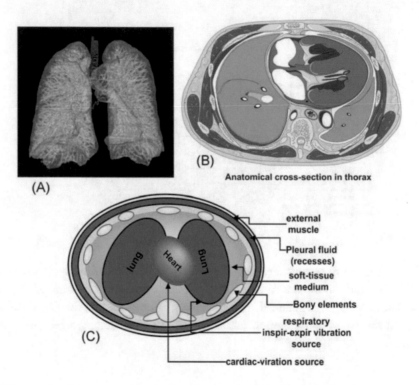

Figure 8.2: Schematic of the cardio-respiratory acoustic model that can be used for signal modeling and identification. This anatomical-based model can be used as calibration reference dynamics, for cardiography acoustic imaging. (A) 3D-reconstruction CT tomography image of the pulmonary alveoli and cardiac bed; (B) transverse cross-section showing the heart muscle (myocardium) and the two lobes of the lung; (C) equivalent cardio-respiratory acoustic model.

Acoustic properties of the solid components of the thorax, such as the chest wall and the heart, are relatively well known. Sound speeds in these tissues are approximately 1,500 m/s, and damping is relatively low. In the larger airways (i.e., diameter of approximately 0.8 mm) of animal models, sound propagates at speeds (mean 95% confidence limit) of 268 ± 44 m/s the acoustic properties of the lung parenchyma, which fills a substantial portion of the human thorax, is a function of the air content of the lung. Parenchymal sound speed was estimated to be relatively low, i.e., between 23 m/s and 60 m/s, depending on air content.

The sound speed reaches a minimum at lung densities that are slightly higher than those at resting volume, and increases from this minimal value of approximately 23 m/s for both higher and lower densities. Therefore, under physiologic conditions, the sound speed is slightly higher in the upper parts of the lung and after inspiration. At resting volume, sound speed is likely closer to 30 m/s than to 60 m/s. As previously noted, the damping characteristics of the lung parenchyma, increases with frequency.

At low audible frequencies, for example 400 Hz, damping is estimated to be only from 0.5–1.0 decibel per centimeter(dB/Cm). Aside from these differences in acoustic properties, the geometrical contribution will be significant to the complexity of cardiac acoustics.

High-frequency sounds are known to travel further within the airway-branching structure, while low-frequency sounds appear to exit predominantly in the large airways via wall motion. Reflections, multiple delays, and interference of acoustic signals, as well as a left-to-right asymmetry of thoracic transmission, will also contribute to the complexity of cardio-thoracic acoustic transmission.

8.3   PATI-EXPERIMENTAL SETUP AND SYSTEM PROTOTYPING

The experimental setup for the phonocardiography acoustic imaging is illustrated in Fig. 8.1, where the electrode (cardiac microphones placement) will place in a sensor matrix for picking up the acoustic vibration of the heart mechanical movement and blood flow information accompanied with it. The sensor matrix is composed of 16x8 microphone elements, by which this matrix will cover the thorax (chest region) anatomically. This sensor distribution will make the acoustic detection as high as possible but in addition it will set the slicing approach of acoustic imaging of thorax cavity but with a limited number of slices.

By using simultaneous multi-cardiac sensor recordings of cardiac sounds (PCG-signals) from the chest wall, an acoustic imaging of the chest has recently been investigated for detecting plausible different patterns between healthy individuals and patients [102]. In that study, a new method for acoustic imaging of the cardiac system was developed and evaluated by a physical model of the cardiac-lung compartment as well as experimental data on four subjects and one patient.

The sound speed, sound wavelengths at the frequencies of diagnostic values, and the geometrical properties of the heart were taken into consideration in developing a model for acoustic imaging to provide a spatial representation of the intra-thoracic sounds as opposed to the mapping of sounds on the thoracic surface.

The acoustic imaging model was developed based on the calculation of a 3D acoustic-array data obtained from the acquired [109]. For convenient declaration of the basic assumption to treat the acoustic mapping problem, the following premise can be considered. The acoustic imaging algorithm tests the hypothesis that it contains the only relevant acoustic source.

A hypothetical source signal is calculated by a least-squares estimation (LSE) method to explain a maximum of the signal variance σ in all microphone signals as follows. Let p_i (i=1,...,CM) be the positions of CM cardiac microphones on the thoracic surface and $D_i(t)$ the signals recorded at these microphones, where (t) represents time.

Assuming a uniform sound propagation throughout the thorax anatomical structure (linear wave propagation), sound speed (c), damping factor per unit length (d), the signal $\phi(y,t)$ emitted by this hypothetical source at the spatial location, (y) can be estimated by solving the linear system Equs. (8.1)–(8.3), which illustrates the signal characterization of phonocardiography acoustic imaging:

$$D_1(t - |p_1 - y|/c) = d^{|p_1-y|} + \phi(y,t)/|p_1 - y|^2, \qquad (8.1)$$
$$D_2(t - |p_2 - y|/c) = d^{|p_2-y|} + \phi(y,t)/|p_2 - y|^2, \qquad (8.2)$$
$$D_{cm}(t - |p_{CM} - y|/c) = d^{|p_{CM}-y|} + \phi(y,t)/|p_2 - y|^2. \qquad (8.3)$$

The dynamic range of amplifier in data acquisition module will be in range of (100–1000 Hz) with noise attenuation level less than (-3 dB). Therefore, the gain schedule of the acquisition channel must be set to the stable region of amplification to avoid any superimposed disturbance signals. The amplifier array should be arranged in a phase array module. The pre-filtering stage, followed the amplification stage with a band-pass filtering (BPF) module, to buffer any frequency drift in acoustic sensor detection channel.

The acoustic array processing was constituted of linear signal processing element with adaptive gain amplifier (AGA); further image reconstruction system (IRS) of the detected acoustic signals would be achieved in high-performance computational PC platform with 64-bit AMD® processor and 4048 MB RAM-system.

The image reconstruction workstation operated on Windows® platform. The radon transformation technique is used to reconstruct acoustic energy mapping of phonocardiography signals; however, the main reconstruction algorithm utilizes adaptive beam-forming approach for sped-up signal processing and frame acquisition rate. The image coordinated of the acquired phonocardiography acoustic mapping will be of transverse plane with inverse z-axis model to be analogous to the anatomical myocardial coordinate system.

The processing algorithm of the acquired acoustic signals in imaging approach will fall into two categories. The first is adaptive beam forming algorithm, which is applied to the filtered PCG signal detected from the cardiac microphone. The input PCG signal to the adaptive beam-forming profile will be as follows:

$$X_{det} = \sum_{l=0}^{L} \Psi_l + \sum_{m=0}^{M} E_m, \qquad (8.4)$$

where the output signal from the adaptive beam-forming unit will be as follows:

$$Y_{det} = \sum_{m=0}^{M} \Psi_m + X_{det}. \qquad (8.5)$$

The parameterized coefficients of the vibro-acoustics signal will be stored in filter-array memory for breath of time-delay (Td). Figure 8.2 illustrates the principal signal processing sequences in phonocardiography acoustic imaging approach. The second approach is the online-Radon transformation with application of back projection algorithm.

The block diagram of the phase array processing for the cardiac acoustic imaging is presented in Fig. 8.3 where the adaptive-filtering kernel applied to the preprocessed PCG signals, with active channel scanning and feature-based extraction technique to be embedded in the image-reconstruction system (IRS).

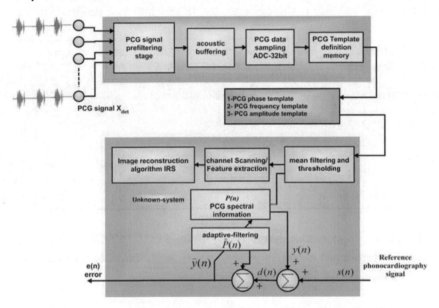

Figure 8.3: Block diagram of adaptive acoustic image reconstruction system.

In this schematic, the phonocardiography signals were multiplexed into a pre-filtering stage, pre-processed through acoustic buffering, and digitally converted using 32-bit sampling resolution.

8.4   ACOUSTIC ARRAY SIGNAL PROCESSING

Various acoustic array configurations have been studied to improve spatial resolution for separating several closely spaced targets and vibration sources in tight formations using unattended acoustic arrays. The application of the acoustic array processing in formation and reconstruction of cardiac acoustic image is still limited in use and needs to improve the received signal quality form the cardiac microphone detectors.

To extend the array aperture as a cardiac detector matrix, it is customary to employ sparse array configurations with uniform inter-array spacing wider than the half-wavelength of phonocardiography signals for intra-subarray spacing, hence achieving more accurate direction of arrival

(DOA) for the PCG signal estimates without using extra hardware configuration. However, this larger inter-array positioning results in ambiguous PCG-DOA estimates.

To resolve this ambiguity, sparse arrays with multiple invariance properties could be deployed. Alternatively, one can design regular or random sparse array configurations that provide frequency diversity, in which case every subarray is designed for a particular band of frequencies.

Additionally, we present a Capon DOA algorithm that exploits the specific geometry of each array configuration. Simulation results are conducted before any realization and implementation of acoustic image system, but it is still prospective imaging modalities to investigate the hemodynamics of the myocardium and vascular pathologies in simple and low-cost profile. Another alternative

Table 8.1: Physiological parameters of cardiac acoustic imaging.

PCG parameters	signal bandwidth (Hz)	maximum amplitude(mV)	minimum amplitude(mV)	attenuation coefficient(α)
S_1	209±1.25	344±1.93	275±1.2	0.238± 0.0043
S_2	245±2.03	302±2.04	207±1.4	0.167± 0.0021
S_3	312±1.69	278±2.07	240±1.75	0.382± 0.0038
S_4	135±1.48	290±2.19	178±1.29	0.246± 0.0031
S_1-S_2	114±1.21	105±1.72	73±1.36	0.129± 0.0027
S_2-S_3	89±1.94	78±2.16	47±1.81	0.203± 0.0030

cardiac acoustic mapping system was developed by Guardo et al. (1998) [103], where he used the posterior microphone sensor pad to rest the patient back (posterior surface of thorax) to simultaneous phonocardiography signal acquisition.

8.4.1   ADAPTIVE BEAM-FORMING IN CARDIAC ACOUSTIC IMAGING

In this section, the spatial form of adaptive signal processing that finds practical use in radar, sonar, communications, geophysical exploration, astrophysical exploration, and biomedical signal processing will be described. In the particular type of spatial filtering of interest to us in this book, a number of independent sensors are placed at different points in space to "listen" to the received signal and in this case it is the phonocardiography signals detected by cardiac microphones.

In effect, the sensors provide means of sampling the received signal in space. The set of sensor outputs collected at a particular instant of time constitutes a snapshot. Thus, a snapshot of data in spatial filtering (for the case when the sensors lie uniformly on a straight line) plays a role analogous to that of a set of consecutive tap inputs that exist in a transversal filter at a particular instant of time. In radar imaging, the sensors consist of antenna elements (e.g., dipoles, horns, slotted waveguides) that respond to incident electromagnetic waves.

In sonar method, the sensors consist of hydrophones designed to respond to acoustic waves. In any event, spatial filtering, known as beam forming, is used in these systems to distinguish between the spatial properties of signal and noise.

The device used to carry out the beam forming is called a beam former. The term beam former is derived from the fact that the early forms of antennas (spatial filters) were designed to form pencil beams, so as to receive a signal radiating from a specific direction and attenuate signals radiating from other directions of no interest [106].

Note that the beam forming applies to the radiation (transmission) or reception of energy. Figure 8.4 shows the spatial mapping of different cardiac-microphone location, by using adaptive PCG beam forming reconstruction, where the spatial-deviation is considered as an binomial function of received microphone intensities. In a primitive type of spatial filtering, known as the delay-and-

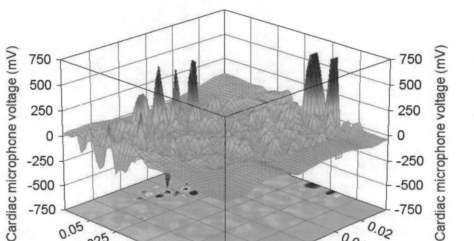

Figure 8.4: Spatial mapping of the PCG acoustic imaging, which assigns the space deviation of PCG signal as a function of output voltage of cardiac microphone.

sum beam former, the various sensor outputs are delayed (by appropriate amounts to align spatial components coming from the direction of a target) and then summed. As Fig. 8.5 illustrates, five microphone locations are specified for covering the all direction-of-arrival (DOA) of the heart acoustic waves. Accordingly, for a single target the average power at the output of the delay-and-sum beam former is maximized when it is steered toward the target. However, a major limitation of the delay-and-sum beam former, even so, is that it has no provisions for dealing with sources of interference.

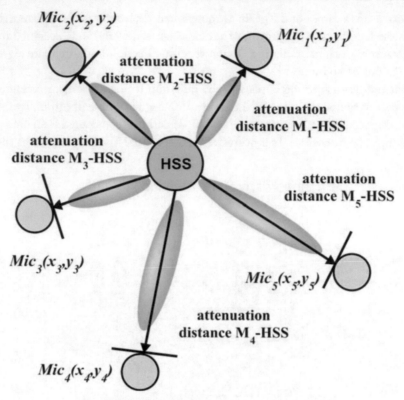

Figure 8.5: Heart sound localization method, where the HSS represents the cardiac auscultation source. The dynamic properties of the heart acoustic vibration, shows a nonlinear behavior, and this is due to the propagation in a non-homogeneous medium through thoracic cavity. Additionally, the curvature of the microphone array is not symmetric due to the geometric parameters of the thorax.

In order to enable a beam former to respond to an unknown interference environment, it has to be made adaptive in such a way that it places nulls in the direction(s) of the source(s) of interference automatically and in real time.

By a similar approach, the output signal-to-noise ratio of the system is increased, and the directional response of the system is thereby improved. This method was also applied in localizing the acoustic source in many engineering and medical applications (e.g., infra-oceanic acoustic mapping and vital signal monitoring based on sonar-wave [103, 105]).

The localization of the acoustic source can be illustrated in a delay-source-receiver method, which is considered as a robust and effective method for heart sound localization. The method is simply based on the delay between signals received at two microphones is found by cross-correlation. For instance, delay1-2 is the delay between the time it takes for the signal to arrive at microphone 1 compared to microphone 2. Then, the delay is converted from sampled time units to distance units, as in Equ. (8.2). This new delay, delta1-2, represents the difference between the distance from microphone (mic1) to the source and from microphone (mic2) to the source (as shown in Fig. 8.5 and equation below):

$$\delta_{mic1-mic2} = (delay_{mic1-mic2}(v_{sound}))/f_s. \tag{8.6}$$

$$\delta_{mic1-mic2} = (dist_{mic1-HSS} - dist_{mic2-HSS}). \tag{8.7}$$

Then, using the standard distance equation, one is able to construct a system of two equations, (8.6) and (8.7), and two unknowns. The coordinates of microphones 1, 2, and 3 are (x_1, y_1), (x_2, y_2), and (x_3, y_3), respectively. The values of these variables are known. The coordinates of the source, (x_s, y_s), are unknown

$$\delta_{mic1-mic2} = \sqrt{(x_1 - x_s)^2 + (y_1 - y_s)^2} - \sqrt{(x_2 - x_s)^2 + (y_2 - y_s)^2} \tag{8.8}$$

$$\delta_{mic1-mic3} = \sqrt{(x_1 - x_s)^2 + (y_1 - y_s)^2} - \sqrt{(x_3 - x_s)^2 + (y_3 - y_s)^2}, \tag{8.9}$$

This two equation, with two unknown variables, which can be solved through any technical computation language such as MATLAB$^{®}$ or LabVIEW$^{®}$ computation software platform.

8.4.2 ADAPTIVE BEAM FORMER WITH MINIMUM-VARIANCE DISTORTIONLESS RESPONSE

Consider the adaptive beam former module that uses a linear array of (M) identical sensors, as presented in Fig. 8.5. The individual sensor outputs, assumed to be in baseband form, are weighted and then summed. The beam former has to satisfy two requirements:

1. A steering capability whereby the target signal is always protected.

2. The effects of sources of interference whereby; the effects are minimized.

One method of providing for these two requirements is to minimize the variance (i.e., average power) of the beam former output, subject to the constraint that during the process of adaptation the weights satisfy the condition. Thus, the target signal steering and source inference will affect the target-image reconstruction with a considerable delay-time [107, 108].

$$w^H(n)s(\phi) = 1 \quad \text{for all n and } \phi = \phi_t, \tag{8.10}$$

where w(n) is the M×1 weight vector, and $s(\phi)$ is an M×1 steering vector. The superscript H denotes Hermitian transposition (i.e., transposition combined with complex signal conjugation). In this application, the baseband data are complex valued, hence the need for complex conjugation.

The value of the electrical angle $\phi = \phi_t$ is determined by the direction of the target. The angle ϕ is itself measured with sensor (1) (at the top end of the array) and treated as the point of reference. The dependence of vector $s(\phi)$ on the angle ϕ is defined by:

$$s(\phi) = [1, e^{-j\phi}, ..., e^{-j(M-1)\phi}]^T. \tag{8.11}$$

The angle ϕ is itself related to incidence angle θ of a plane wave, measured with respect to the normal to the linear array as follows:

$$\phi = \frac{2\pi d}{\lambda} sin(\theta), \tag{8.12}$$

where (d) is the spacing between adjacent sensors of the array and λ is the wavelength. The incidence angle (θ) lies inside the range $-\pi/2$ to $\pi/2$. The permissible values that the angle ϕ may assume lie inside the range ($-\pi$ to π). This means that we must choose the spacing d < $\lambda/2$, so that there is a one-to-one correspondence between the values of ϕ and θ without ambiguity. The condition d < $\lambda/2$ may be viewed as the spatial analog of the sampling theorem.

Figure 8.6 displays the general schematic of PATI-imaging technique, where (A) cylindrical model of cardiac microphone positions, (B) a corresponding chest x-ray radiography shows, the planar view of thorax anatomy, and (C) reconstructed acoustic mapping for microphones intensity and the spatial location of them.

8.4.3   HEART SOUNDS PHYSIOLOGICAL MODELING BASED ON PATI METHOD

The present interaction between medical imaging and modeling is mutual validation-the process of comparing data from model and imaging systems. This integration has numerous advantages from both modeling and imaging perspectives. Firstly, validating model predictions against imaging data provides a mechanism for testing that the model captures all the key physiological components of the system.

This is performed for a prescribed set of parameter values [110, 112], and once completed, the model is a powerful tool to establish predictions of the system properties in new regimes.

From an imaging perspective, models can be used to extract information that is not directly available from the images themselves and thus will assist in the clinical diagnosis. For example, the mechanical stress or work of a contractile tissue cannot be detected directly from an image but is straightforward to extract from a model parameterized from same information.

This simulation-based imaging (in silico) approach provides significant capacity to define new metrics for focusing clinical trials [112, 115, 116], optimizing patient selection and customizing therapy [106, 107]. The relationship between the new invited modality of (PATI) as cardiac acoustic mapping technique and relevant physiological modeling approach for the cardiac hemodynamics is mutually connected in its nature and have consisted application in physio-acoustic system identification.

The use of PATI technique will be such a prospective research in modeling and simulation cardiac acoustic dynamics in accompanied with multi-modality medical imaging such as high-resolution computerized tomography (Hi-Rez CT). Figure 8.7 shows the image reconstruction of various microphone located spatially, which detects the defined target acoustic variation in a geometric setting.

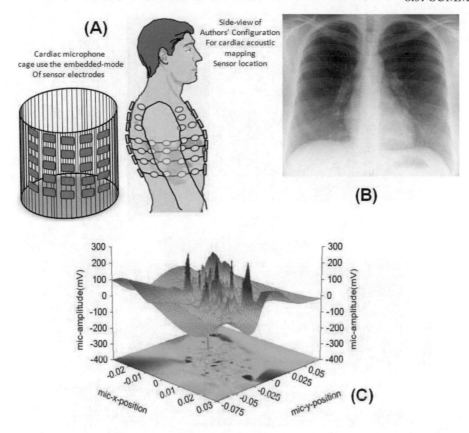

Figure 8.6: General cardiac acoustic imaging platform and its reconstructed acoustic mapping features acquired simultaneously with ECG gated signal. (A) is the geometrical representation of the cardiac microphone array; (B) is the AP projection chest radiography demonstrating the main anatomical parts involved in acoustics imaging; (C) is the spatial intensity mapping for x-positioned microphone and y-positioned microphone of dummy load model for thorax anatomy.

The advances in the real-time image reconstruction [111, 113, 114, 117] and image acquisition optimization will also reflect on the further development of the precise and accurate cardiac acoustic imaging [108, 110].

8.5   SUMMARY

The cardiac acoustic imaging method is considerably non-competitive to other medical imaging modalities such as the x-ray, CT, MRI, and PET-imaging techniques in terms of the anatomical and physiological information that they supply. The preliminary experimental results are encour-

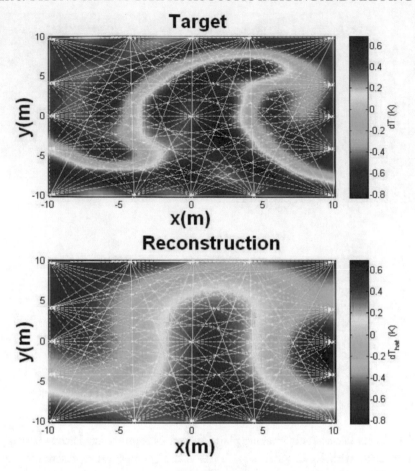

Figure 8.7: The dynamic variation of the detectable acoustic waves in the dummy-phantom target for testing the performance of the microphone array detector in spatial domain as vertical positioned microphone y(m) and horizontal positioned x(m)(above)moving targets with 0.0023 m/s (below) image reconstruction.

aging and motivates researcher to be open to a new diagnostic possibility. To stand that further optimization and development in cardiac microphone sensor technology, hardware design, signal processing algorithms, and advances in information technology to reduce the computational cost of the algorithm, the PATI-imaging technique may become a routine monitoring technique before using the other conventional methods, due to its noninvasive approach and simple implementation method.

In addition, the technical complications and the deficiency in the cardiac acoustic image quality can also be improve by synchronized compensation of additive noise and acoustic disturbances which may infer spatial and temporal representation of the detected PCG signals.

Although this new medical imaging suffers from a low spatial and temporal resolution, it could be proved to be a good choice for low-cost and mobility strategy in cardiac imaging, rather than the ultrasound imaging and SPECT imaging module. The expected research direction should be guided to improve the image quality, increase the SNR value of the acoustic detectors, and enhance the microphone-matrix design to be a good choice for clinical imaging application.

Feedback from Clinical and Biomedical Aspects

The new direction of medical engineering toward more automated and intelligent systems reflects on the future trends of many clinical applications of biomedical signal processing, and this indeed will optimize both the diagnostic methods and the philosophy of clinical and medical data interpretation-analysis loop. Moreover, the use of the new embedded technology, artificial intelligence, higher computational algorithms, and state-of-the-art biomedical devices will make the health care process and its collaterals more effective in assigning, integrating and delivering the suitable medical care to the vast number of patients.

In addition, the rapid and expected needs in developing and enhancing the quality of life will lead to increasing expectations for the inventing and synthesis of new medical technologies originated from current trends in development and practical application of the life sciences. The development of non-invasive biomedical instrumentation technology to measure and transduce the physiological information and signals generated by the living subject in a minimally invasive and continuous fashion, was responsible for the establishment of the reasonable concept of patient monitoring. Therefore, the impression of cardiac acoustic signal signature on the clinical and medical practicing will increase in its impression, and will be of great interest for biomedical and clinical engineers who work in the physiological and bioacoustics field.

Although this field has many technical and physical obstacles, but it is considered a straight-forward and easy-to-handle medical diagnostic procedure. The phonocardiography and auscultation technique is still nascent, compared to advanced analysis methods. On the other hand, the slow development curve of cardiac-acoustics analysis itself provides an open field for promoting research and investigation. The advanced technique of heart sound analysis and processing should be more impressive and robust, in order to make the cardiac auscultation method more productive in clinical diagnosis and medical primary care. The combination of multiple points of view, from clinical and engineering specialists, will be of a great influence on maturity of biomedical instrumentation and medical technology.

This book, from the authros' point of view, provides a good reference for the biomedical engineering student who seeks productive resources in phonocardiography signal processing and its related fields of interests. The book also presents a relatively new approach in cardiac acoustic mapping, which seems to be of a great privilege for advanced research in bioacoustics. Additionally, this will require appropriate post-processing techniques that efficiently deal with phonocardiography signal.

Bibliography

[1] A. A. Luisada, *From Auscultation to Phonocardiography*, Mosby, Saint Louis, 1965.

[2] C. Ludeman Lonnie, *Fundamentals of digital signal processing*, Harper and Row Publishers, Inc. 1986.

[3] K. Vinay Angle and G. John Proakis, *Digital Signal Processing Using Matlab V4*, PSW publishing company, 1997.

[4] Ashok Ambardar, *Analog and digital signal processing*, PSW publishing company, 1995.

[5] M. Kuo Sin and H. Bob Lee, *Real time signal processing: Implementation, Applications, and Experiments with TMS320C55X*, John Wiley &Sons, LTD, 2001.

[6] Stergiopoulos, Stergios, *Advanced Signal Processing Handbook: Theory and Implementation of Radar, Sonar, and Medical Imaging Real-Time Systems*, Boca Raton: CRC Press LLC, 2001.

[7] W. C. Knight, R. G. Pridham, and S.M. Kay, Digital signal processing for sonar, *Proc. IEEE*, 69(11), 1451-1506, 1981. DOI: 10.1109/PROC.1981.12186

[8] B. Windrow and S. D. Stearns, *Adaptive Signal Processing*, Prentice-Hall, Englewood Cliffs, NJ, 1985.

[9] A. B. Baggeroer, *Sonar signal processing, in Applications of Digital Signal Processing*, A.V. Oppenheim, Ed., Prentice-Hall, Englewood Cliffs, NJ, 1978.

[10] H. J. Scudder, Introduction to computer aided tomography, *Proc. IEEE*, 66(6), 628-637, 1978. DOI: 10.1109/PROC.1978.10990

[11] D. Nahamoo and A. C. Kak, *Ultrasonic Diffraction Imagin*, PTR-EE 82-80, Department of Electrical Engineering, Purdue University, West Lafayette, IN, August 1982.

[12] S. W. Flax and Mc. Donnell, Phase-aberration correction using signals from point reflectors and diffuse scatterers: basic principles, *IEEE Trans. Ultrasonic, Ferroelectric Frequency Control*, 5(6), 758-767, 1988. DOI: 10.1109/58.9333

[13] G. C. Ng, S. S. Worrell, P. D. Freiburger, and G. E. Trahey, A comparative evaluation of several algorithms for phase aberration correction, *IEEE Trans. Ultrasonic, Ferroelectrics Frequency Control*, 1(5), 631-643, 1994. DOI: 10.1109/58.308498

[14] A. K. Jain, *Fundamentals of Digital Image Processing*, Prentice-Hall, Englewood Cliffs, NJ, 1990.

[15] Q. S. Xiang and R.M. Henkelman, K-space description for the imaging of dynamic objects, *Magn. Reson. Med.*, 29, 422-428, 1993. DOI: 10.1002/mrm.1910290324

[16] M. L. Lauzon, D.W. Holdsworth, R. Frayne, and B.K. Rutt, Effects of physiologic waveform Variability in triggered MR imaging: theoretical analysis, *J. Magn. Reson. Imaging*, 4(6), 853-867, 1994. DOI: 10.1002/jmri.1880040618

[17] C. J. Ritchie, C. R. Crawford, J. D. Godwin, K. F. King, and Y. Kim, Correction of computed Tomography motion artifacts using pixel-specific back-projection, *IEEE Trans. Medical Imaging*, 15(3),333-342, 1996. DOI: 10.1109/42.500142

[18] A. Leatham, *Auscultation of the Heart and Phonocardiography*. Churchill Livingstone, London, Edinburgh, 1975.

[19] J. Elizabeth Corwin, *Handbook of Pathophysiology: Foundations of Health and Disease*, 3rd Ed. 2006.

[20] C. Guyton Arthur and E. Hall John, *Textbook of Medical Physiology*, Saunders W.B., 2005.

[21] Poul-Erik Paulev, *Textbook in Medical Physiology and Pathophysiology Essentials and clinical problems*, Copenhagen Medical Publishers, 2000.

[22] T.A.D Michael, *Auscultation of the Heart*. McGraw-Hill, New York, 1997.

[23] B. Erickson, *Heart Sounds and Murmurs: A Practical Guide*. Mosby-Year Book, Inc., St. Louis, 1997.

[24] M. S. Obaidat, Phonocardiogram signal analysis: techniques and performance comparison, *MS Obaidat-Journal of medical engineering and technology*, Nov-Dec 17(6):221-7, 1993. DOI: 10.3109/03091909309006329

[25] H. Liang, S. Lukkarinen and I. Hartimo, Heart sound segmentation algorithm based on heart sound envelogram, *Computational Cardiology*, v24, 105-108, 1999. DOI: 10.1109/CIC.1997.647841

[26] A. C. Kak and M. Slaney, and G. Wang, Principles of computerized tomography imaging, *Medical Physics, IEEE Press*, New York, 1992.

[27] N. C. Nanda, *Doppler Echocardiography*, 2ed, Lea and Febiger Publisher, London, 1993.

[28] Bill Gardner and Keith Martin, *HRTF Measurements of a KEMAR Dummy-Head Microphone*, MIT Media, Aug, 2000.

[29] E. Mannheimer, Standardization of phonocardiography, *Proc. of Second European Congress of Cardiology*, 1956.

[30] B. G. Wells, Rappaport M. B., Sprague H. B., The sounds and murmurs in contraction of the aorta: a study by auscultation and phonocardiography, *Am. Heart J.*, Jul. 38(1):69-79, 1949. DOI: 10.1016/0002-8703(49)90793-0

[31] K. Holldack, Clinical significance of auscultation using electronic amplifiers, *Zeitschrift. Gesamte. Inn Med.*, Oct 15, 28(20):606-8. Germany, 1973.

[32] H. Ljunggren, Investigation of an apparatus for calibrated phonocardiography according to the Mannheimer-Stordal system, *Acta Med Scand.*, Jun 30, 133(6):388-93, 1949.

[33] D. H. Bekkering, Weber J., Standardization of phonocardiography: efforts in the Netherlands, *Am. Heart J.*, Aug, 54(2):316-7, 1957. DOI: 10.1016/0002-8703(57)90162-X

[34] A. Leatham, Auscultation and phonocardiography: a personal view of the past 40 years, *Br. Heart J.*, May, 57(5):397-403, 1987. DOI: 10.1136/hrt.57.5.397

[35] A. A. Luisada, *From Auscultation to Phonocardiography*, Mosby, Saint Louis, 1965.

[36] R. Zalter, H. C. Hardy, A. A. Luisada, Acoustic transmission characteristics of the thorax, *J Appl Physiol.*, Mar, 18:428-36, 1963.

[37] O. Zoneraich and S. Zoneraich, *Phonocardiographic Findings in Patients with Malfunctioning Artificial Valves, in Non-Invasive Methods in Cardiology*, S. Zoneraich, Ed. Springfield, II: Charles C Thomas, 1974.

[38] S. G. Kahalekar, S. J. Vaidya, A. Shrawagi, PC based phonocardiograph expert system, *Proceedings RC IEEE-EMBS and 14th BMESI*, vol. 2, New Delhi, India, pp. 64-65, 1995.

[39] R. Zalter, H. Hodara, A. A. Luisada, Phonocardiography. I. General principles and problems of standardization., *Am. J Cardiol.*, Jul, 4(1):3-15, 1959. DOI: 10.1016/0002-9149(59)90188-2

[40] S. Pavlopoulos, A. Stasis, E. Loukis, A decision tree-based method for the differential diagnosis of aortic stenosis from mitral regurgitation using heart sounds, *Biomedical Engineering Online*, 3:21. 2004. DOI: 10.1186/1475-925X-3-21

[41] P. D. Stein, H. N. Sabbah, J. B. Lakier, and S. Goldstein, Frequency spectrum of the aortic component of the second heart sound in patients with normal valves, aortic stenosis and aortic porcine xenografts, *Amer. J. Cardiology.*, vol. 46, pp. 48-52, July 1980. DOI: 10.1016/0002-9149(80)90604-9

[42] B. Kingsley, Acoustic evaluation of prosthetic cardiac valve in the audio spectrum, *J. Audio Eng. Soc.*, pp. 750-755, Nov. 1972,

[43] R. F. Gordon, M. Najmi, B. Kingsley, B. L. Segal, and J. Linhart, Spectroanalytic evaluation of aortic prosthetic valves, *Chest*, vol. 66, pp. 44-49, July 1974. DOI: 10.1378/chest.66.1.44

[44] S. L. Xu, M. Hu, D. C. Zhang, The time-frequency representation of PCG, Industrial technology, *Proceedings of the IEEE International Conference*, Los Angeles, USA, pp. 679-681, 1994. DOI: 10.1109/ICIT.1994.467056

[45] P. D. Stein, H. N. Sabbah, J. B. Lakier, D. J. Magilligan, Jr., and S. Goldstein, Frequency of the first heart sound in the assessment of stiffening of mitral bioprosthetic valves, *Circulation*, vol. 63, pp. 200-203, 1981.

[46] J. P. Burg, Maximum entropy spectral analysis, presented at the *P37th S.E.G. Annual Meeting*, Oklahoma City, OK, 1967,

[47] S. S. Kraman, Speed of low-frequency sound through the lungs of normal men, *J. Appl. Physiol.*, Vol. 55, pp. 1862–1867, 1983.

[48] T. H. Joo, *Pole-zero modeling and classification of phonocardiograms*, M.Sc. thesis, Dept. Electrical Engineering and Computational Science, Massachusetts Institute of Technology, Cambridge, MA, Jan. 1982.

[49] D. Rushmer, *Cardiovascular Dynamics*, Sounders publisher, Philadelphia, 1976.

[50] M. L. Jacobson, Analysis and classification of physiological signals using wavelet transforms, *Proceedings of the 10th IEEE International Conference on Electronics, Circuits and Systems*, Dec, 2003. DOI: 10.1109/ICECS.2003.1301934

[51] M. Kompis, H. Paskterkamp and G. Wodicka, Acoustic imaging of the human chest, *J. Chest*, Vol. 120, No. 4, pp. 1309–1321, 2001. DOI: 10.1378/chest.120.4.1309

[52] R. C. Hilborn, *Chaos and Nonlinear Dynamics*, 2nd edition, Oxford, Oxford University Press, 2000.

[53] Åström, K. J. and P. Eykhoff. System identification— a survey, *Automatica*, 7, 123–162. 1971. DOI: 10.1016/0005-1098(71)90059-8

[54] P. Soulie, Intra-cardiac phonocardiography, *III-World Congress of Cardiology*, Washington, 1954.

[55] J. T. Willerson, J. A. Kastor, R. E. Dinsmore, E. Mundth, M. J. Buckley, W. Gerald Austen, and C. Sanders, Non-invasive assessment of prosthetic mitral paravalvular and intravalvular regurgitation, *Brit. Heart J.*, vol. 34, pp. 561-568, 1972. DOI: 10.1136/hrt.34.6.561

[56] T. Kohonen, Self-organized formation of topologically correct feature maps, *Biological Cybernetics.*, 43(1):59–69, 1982. DOI: 10.1007/BF00337288

[57] C. L. Nikias and A. P. Petropulu, *Higher-Order Spectra analysis*, PTR Prentice Hall, New Jersey, 1993.

[58] Brockway, Marina, et al., Method and Apparatus for Optimization of Cardiac Resynchronization Therapy Using Heart Sound, *PU.S. Appl. No. 10/865,498,* filed Jun. 10, page 45, 2004.

[59] P. D. Stein, H. N. Sabbah, D. T. Anbe, F. Khaja, G. M. Folger, Diagnostic. Value of intra-cardiac phonocardiography in subaortic stenosis, (abstract) *Chest,* 72:413, 1977,

[60] A. Wennevold, Right-heart intra-cardiac phonocardiography in 80 subjects without organic heart disease: A study of the origin of innocent heart murmurs, *Dan Med Bull 15: 313,* 1968.

[61] P. Boutouyrie, S. Laurent, X. Girerd, A. Benetos, P. La Colley, E. Abergel, M. Safar, Common carotid artery stiffness and patterns of left ventricular hypertrophy in hypertensive patients. Hypertension, *25: 651–659,* 1995.

[62] E. Craige, On the genesis of heart sounds. Circulation, *1976; 53:207–209,*

[63] Silverman M., A. Leatham, Twentieth century pioneer in auscultation, *Clinical Cardiology,* 155–157, 1999.

[64] M. II. Drasncr, J. E. Rame, L. W. Stevenson, and D. L. Dries, Prognostic importance of elevated jugular venous pressure and a third heart sound in patients with heart failure. *N Engl J Med,* 345:574–581, 2001. DOI: 10.1056/NEJMoa010641

[65] C. F. Wooley, P. B. Baker, A. J. Kolibash, J. W. Kilman, E. A. Sparks, H. Boudoulas, The floppy, myxomatous mitral valve, mitral valve prolapse, and mitral regurgitation. *Prog. Cardiovascular Disease,* 33:397–433, 1991. DOI: 10.1016/0033-0620(91)90005-7

[66] D. Braschdorff, S. Ester, T. Dorsel, and E. Most, Neural network based multi sensor heart sound analysis, *Proc. Comput. Cardio.* pp. 303-306, 1991. DOI: 10.1109/CIC.1990.144221

[67] A. K. Abbas, Rasha Bassam, Adaptive ARMAX Cardiac Acoustic Identification, International Conference on Modeling, Simulation and Optimization ICMSAO 2009, Al Sharjah-UAE, *an,* pp 209-213, 2009.

[68] *Sound and vibration analysis,* theoretical basis for software development, User Guide(CHML) National Instruments (www.ni.com),

[69] P. P. Pinna, E. Piccolo, S. Bartolozzi and V. Fontana, The FFT in the study of the fourth heart sound. IEEE Computational Cardiology, *369-372,*

[70] J. P. Burg, Maximum entropy spectral analysis, presented at the 3 *th S.E.G. Annual Meeting,* Oklahoma City, OK, 1967.

[71] L. Khadra, M. Matalgah, B. El-Asir and S. Mawagdeh, The wavelet transform and its application to phonocardiogram signal analysis, *Med. Inf. v6. 271-277,* DOI: 10.3109/14639239109025301

[72] M. Matalgah, J. Knopp, S. Mawagdeh, Interactive processing method using Gabor wavelet and the wavelet transform for the analysis of phonocardiogram signals, *Time-Frequency and Wavelets in Biomedical Signal Processing,* IEEE Engineering in Medicine and Biology Society. pp. 271-304, 1997.

[73] H. P. Sava, R. Bedi, J. T. E. McDonnell, P. M. Grant, Classification of Carpentimer-Edwards bioprosthesis heart valves using an adaptive single layer perceptro, *Proceedings of Engineering in Medicine and Biology Society, Montréal,* CA, Vol.1, pp. 129-130, 1995. DOI: 10.1109/IEMBS.1995.575034

[74] J. E. Hebden, J. N. Torry, Neural network and conventional classifiers to distinguish between first and second heart sounds. Artificial Intelligence Methods for Biomedical Data Processing, *IEEE Colloquium,* April, London, vol.3, pp. 1-6, 1996.

[75] M. Akay, Y. M. Akay, W. Welkowitz, Automated noninvasive detection of coronary artery disease using wavelet-based neural networks, *Proceedings of the 16th Annual International Conference of Engineering in Medicine and Biology Society,* Baltimore, USA, vol. 1, pp. A12-A13, 1994. DOI: 10.1109/IEMBS.1994.412126

[76] N. Levanon and E. Mozeson, *Radar Signals,* J. Wiley and Sons (Interscience Div.) New York, 2004. DOI: 10.1002/0471663085

[77] P. Wang, C.S. Lim, S. Chauhan, Foo, J.Y.A., and Anantharaman, V., Phonocardiographic Signal Analysis Using a Modified Hidden Markov Model. Ann. Biomedical Eng., *v35 i3. 367-374,* DOI: 10.1007/s10439-006-9232-3

[78] S. Molau, M. Pitz, R. Schluter, H. Ney, Computing Mel-frequency cepstral coefficients on the power spectrum, IEEE International Conference on Acoustics, Speech, and Signal Processing, *ay 7-11,* Utah, USA, pp. 73-76, 2001. DOI: 10.1109/ICASSP.2001.940770

[79] D. Braschdorff, S, Ester, T. Dorsel, and E. Most, Neural network based multi-sensor heart sound analysis, *Proc. Comput. Cardio.,* pp. 303-306, 1991. DOI: 10.1109/CIC.1990.144221

[80] L. R. Rabiner, A tutorial on Hidden Markov Models (HMM) and selected applications in speech recognition, *Proceeding,* IEEE. 257-286,

[81] X. Li, M. Parizeau and R. Plamondon, Training hidden Markov models with multiple observations-a combinatorial method, *IEEE Trans. PAMI. v22. 371-377,* DOI: 10.1109/34.845379

[82] L. R. Rabiner, R. W. Schafer, *Digital Processing of Speech Signals*, Georgia Institute of Technology, 1978.

[83] W. E. Daniel Schobben and C. W. Piet Sommen, A New Algorithm for Joint Blind Signal Separation and Acoustic Echo Canceling, F *fth International Symposium on Signal Processing and its Applications*, ISSPA'99, Brisbane, Australia, 22-25 August, 1999. DOI: 10.1109/ISSPA.1999.815814

[84] T. Leung, P. White, W. Collis, E. Brown and A. Salmon, Classification of heart sounds using time-frequency method and artificial neural networks, *in Proceedings of the 22nd Annual International Conference of the IEEE Engineering in Medicine and Biology Society,* vol.2. pp. 988-991, DOI: 10.1109/IEMBS.2000.897889

[85] I.T. Jolliffe, *Principal Component Analysis,* Springer Series in Statistics, 2nd ed., Springer, NY, 2002.

[86] R. J. Rangayyan, R. J. Lehner, *Phonocardiogram Signal Analysis: a Review,* Durand L.G., Pibarot P.,(Digital signal processing of the phonocardiogram: Review of the most recent advances), CRC Critical Reviews in Biomedical Engineering, vol. 15, pp. 211-236, 1988.

[87] J. C. Dunn, A fuzzy relative of the ISODATA process and its use in detecting compact well-separated clusters, *Journal of Cybernetics,* 3: 32-57, 1973. DOI: 10.1080/01969727308546046

[88] C. James Bezdek, J. Richard Hathaway, Christopher Leckie, Kotagiri Ramamohanarao, RBApproximate data mining in very large relational data. *ADC,* 3-13, 2003.

[89] T. Kohonen, Self-organized formation of topologically correct feature maps, *Biological Cybern.,* 43(1):59–69, 1982. DOI: 10.1007/BF00337288

[90] S. P. Luttrell, Self-organization: A derivation from first principles of a class of learning algorithms, *IJCNN International Joint Conference on Neural Networks,* 1989. DOI: 10.1109/IJCNN.1989.118288

[91] E. Gmachl, Clinical use of phonocardiography, *Wiener klinische Wochenschrift 62(44):836,* Nov 3, 1950.

[92] Ej Ellis, Oh Gauer, Eh Wood, An Intracardiac Manometer: Its Evaluation And Application, *Circulation,* 1951.

[93] G. A. Feruglio, A new method for producing, calibrating, and recording intracardiac sounds in man, *Am Heart J.,* Mar, 65:377-90, 1963. DOI: 10.1016/0002-8703(63)90013-9

[94] P. Soulie, Intra-cardiac phonocardiography, *III World Congress of Cardiology,* Washington, 1954.

[95] Widrow, B. Glover, J.R., Jr. McCool, J.M. Kaunitz, J. Williams, C.S. Hearn, R.H. Zeidler, J.R. Eugene Dong, Jr. Goodlin, R.C., Adaptive noise cancelling: Principles and applications, *IEEE Proceeding*, 1975. DOI: 10.1109/PROC.1975.10036

[96] P. D. Stein, H. N. Sabbah, D. T. Anbe, F. Khaja, G. M. Folger, Diagnostic: value of intra-cardiac phonocardiography in subaortic stenosis, (abstract) *Chest 72: 413*, 1977.

[97] L. G. Gamero and R. Watrous, Detection of the first and second heart sound using probabilistic models, Engineering in Medicine and Biology Society Proc. *5th Annual International Conference of the IEEE* vol.3, pp 2877–80, 2003. DOI: 10.1109/IEMBS.2003.1280519

[98] N. T. Gordienko, N. I. Berseneva, Masterkovana, A. B. Orin, V. A. Silin, Method and apparatus for intracardiac phonography, *Med Prom SSSR.*, Feb, 17:49-52, 1963.

[99] Brockway, Marina, et al., Method and Apparatus for Optimization of Cardiac Resynchronization Therapy Using Heart Sound, *PU.S. Appl. No. 10-865,498*, filed Jun. 10, page 45, 2004.

[100] A. Haghighi-Mood and J. N. Torry, A sub-band energy tracking algorithm for heart sound segmentation, *Computational Cardiology*, pp 22-50, 1995. DOI: 10.1109/CIC.1995.482711

[101] Rasha Bassam, K. Abbas Abbas, M. Kasim Rana, PCG spectral Estimation based on ARMAX technique, 4th Kuala Lumpur International Conference on Biomedical Engineering *BIOMED 2008*, Kuala Lumpur, Malaysia, 2008.

[102] S. S. Kraman, Speed of low-frequency sound through the lungs of normal men, *Journal. Appl. Physiol.*, Vol. 55, pp. 1862–1867, 1983.

[103] M. Cozic, LG Durand, R Guardo, Development of a cardiac acoustic mapping system, *Journal of Medical and Biological Engineering and Computing*, 1998. DOI: 10.1007/BF02523210

[104] M. Kompis, H. Paskterkamp and G. Wodicka, Acoustic imaging of the human chest, *J. Chest*, Vol. 120, No.4, pp. 1309–1321, 2001. DOI: 10.1378/chest.120.4.1309

[105] S. Haykin, Adaptive filters, *Signal Processing Magazine, IEEE Computer Society*, pp. 386-408, 1999.

[106] B. Van Veen and K. Buckley, Beamforming: A versatile approach to spatial filtering, *IEEE ASSP Magazine*, vol. 5, pp. 4-24, April 1988. DOI: 10.1109/53.665

[107] R. C. Hilborn, *Chaos and Nonlinear Dynamics*, 2nd edition, Oxford, Oxford University Press, 2000.

[108] O. T. Von Ramm, S. W. Smith, Beam steering with linear arrays, *IEEE Transactions on Biomedical Engineering*, 1983. DOI: 10.1109/TBME.1983.325149

[109] S. M. Debbal and F. Bereksi-Reguig, Analysis and study of the variation of splitting In the second heart beat sound of the wavelet transform, *Journal of Medical Engineering and Technology*, 2005.

[110] E. Craige, On the genesis of heart sounds, *Circulation*, 53:207–209, 1976.

[111] M. Silverman, A. Leatham, Twentieth century pioneer in auscultation, *Clinical Cardiology*, 22:155–157, 1999. DOI: 10.1002/clc.4960220223

[112] P. Boutouyrie, S. Laurent, X. Girerd, A. Benetos, P. Lacolley, E. Abergel, M. Safar, Common carotid artery stiffness and patterns of left ventricular hypertrophy in hypertensive patients, *Hypertension*, 25: 651–659, 1995.

[113] L. G. Durand and Pibarot P., Digital signal processing of the phonocardiogram: review of the most recent advancements, *Crit. Rev. Biomedical Engineering*, V 23, 163-219, 1995.

[114] H. A. McCann, J. C. Sharp, T. M. Kinter, C. N. McEwan, C. Barillot, J. F. Greenleaf, Multidimensional ultrasonic imaging for cardiology, *Proceedings of the IEEE*, Vol. 79, pp 1063-1073, 1988. DOI: 10.1109/5.9652

[115] T Sakamoto, R Kusukawa, Dm Maccanon, A. A. Luisada, Hemodynamic determinants of the amplitude of the first heart sound, *Circulation Research*, 1965.

[116] C. F. Wooley, P. B. Baker, A. J. Kolibash, J. W. Kilman, E. A. Sparks, H. Boudoulas, The floppy, myxomatous mitral valve, mitral valve prolapse, and mitral regurgitation, *Prog. Cardiovascular Disease*, 33:397–433, 1991. DOI: 10.1016/0033-0620(91)90005-7

[117] D. Rushmer, *Cardiovascular Dynamics*, Sounders Publisher, Philadelphia, 1976.

Contributors

ASSIST. PROF. DR. NOAMAN M. NOAMAN

Noaman M. Noaman received his B.Sc. degree, M.Sc. degree, and Ph.D. degree in Computer and Control Engineering, University of Technology, Baghdad, Iraq in 1983, 1990, and 2001, respectively. He is currently the Head of Electronics Engineering Department at Computer Man College for computer studies, Khartoum, Sudan. Dr. Noaman has published many journal articles and spoke at numerous scientific conferences. His current research interests are in intelligent control system and biomedical engineering. He teaches undergraduate and graduate courses in signal processing and intelligent control engineering.

PROF. DR. FAKER SALMAN AL-ANI

Faker Salman Al-Ani received his Ph.D. degree in Neurophysiology from University of Baghdad, Iraq, 1996, and was the head of the department of Medical Physiology at Nahrain College of Medicine, Baghdad. He is currently a vice dean of faculty of Medicine at Mutah University-Jordan, a potition he has held since 2008, and has published many papers on neurophysiology and clinical physiology. His special interests are Neurophysiology, EEG, EMG, ER, ERG.

About the Authors

ABBAS KHUDAIR ABBAS AL-ZUBAIDY

Abbas Khudair Abbas Al-Zubaidy, was born in Baghdad, Iraq, in 1979. He received his B.Sc. degree in Medical Engineering in 2001 from Saddam University, Baghdad, and his M.Sc. degree in Cardio-vascular Engineering from the same university in 2004. He has been a DAAD (German Academic Exchange Service) Scholarship holder since 2006 in a doctoral study at the Helmholtz Institute of Biomedical Engineering, RWTH Aachen University-Germany. Currently he is doing his Ph.D. project toward development of non-invasive temperature monitoring based on infrared thermography (IRT) imaging technique, for the prediction of physiological and clinical status of the neonate in response to cold and hot stress. He is interested in biomedical signal processing and physiological data clustering techniques. He has published more than 20 conference and journal papers in the field of mechatronic application and biomedical signal processing in clinical analysis module. Formerly he worked as an application engineer in Philips Medical System in Iraq from 2004-2006, he first applied MRS spectroscopy signal acquisition in Iraq, at Al-Khadumiya Teaching Hospital-Nahrain University as a collaboration with the Neurosurgery Department for the development of the lesions localization method based on the MRS-imaging application.

RASHA BASSAM

Rasha Bassam was Born in Baghdad, Iraq in 1984. She received her B.Sc. degree with honors at the Medical Engineering Department of Saddam University in 2005, in Baghdad. She then got a DAAD (German Academic Exchange Service) scholarship to study for her M.Sc. degree in biomedical engineering. She received her M.Sc. degree in biomedical engineering at Aachen University of Applied Sciences in 2009. Currently she is also a DAAD scholarship holder for a doctoral study at the Bioengineering Institute at Aachen University of Applied Sciences, Juelich. Germany. She is interested in cellular biophysics, protein dynamic obscurities, water problems, nitric oxide mysteries, biomedical signal processing, physiological data classification, and clustering techniques. She has published about 15 conference, research review, and journal papers in the fields of protein biophysical dynamics, mechatronic application, and biomedical signal processing in clinical analysis module.

Printed in the United States
by Baker & Taylor Publisher Services